W9-BCL-831

Towards a Sustainable University

DOI: 10.1057/9781137351937

Other Palgrave Pivot titles

Jordi Cat: **Maxwell, Sutton and the Birth of Color Photography: A Binocular Study**

Nevenko Bartulin: **Honorary Aryans: National–Racial Identity and Protected Jews in the Independent State of Croatia**

Coreen Davis: **State Terrorism and Post-transitional Justice in Argentina: An Analysis of Mega Cause I Trial**

Deborah Lupton: **The Social Worlds of the Unborn**

Shelly McKeown: **Identity, Segregation and Peace-Building in Northern Ireland: A Social Psychological Perspective**

Patrick Joseph Ryan: **Master-Servant Childhood: A History of the Idea of Childhood in Medieval English Culture**

Rita Sakr: **'Anticipating' the 2011 Arab Uprisings: Revolutionary Literatures and Political Geographies**

Timothy Jenkins: **Of Flying Saucers and Social Scientists: A Re-Reading of When Prophecy Fails and of Cognitive Dissonance**

Ben Railton: **The Chinese Exclusion Act: What It Can Teach Us about America**

Andrew Dowdle, Scott Limbocker, Song Yang, Karen Sebold, and Patrick A. Stewart: **Invisible Hands of Political Parties in Presidential Elections: Party Activists and Political Aggregation from 2004 to 2012**

Jean-Paul Gagnon: **Evolutionary Basic Democracy: A Critical Overture**

Mark Casson and Catherine Casson: **The Entrepreneur in History: From Medieval Merchant to Modern Business Leader**

Tracy Shilcutt: **Infantry Combat Medics in Europe, 1944–45**

Asoka Bandarage: **Sustainability and Well-Being: The Middle Path to Environment, Society, and the Economy**

Panos Mourdoukoutas: **Intelligent Investing in Irrational Markets**

Jane Wong Yeang Chui: **Affirming the Absurd in Harold Pinter**

Carol L. Sherman: **Reading Olympe de Gouges**

Elana Wilson Rowe: **Russian Climate Politics: When Science Meets Policy**

Joe Atikian: **Industrial Shift: The Structure of the New World Economy**

Tore Bjørgo: **Strategies for Preventing Terrorism**

Kevin J. Burke, Brian S. Collier and Maria K. McKenna: **College Student Voices on Educational Reform: Challenging and Changing Conversations**

Raphael Sassower: **Digital Exposure: Postmodern Postcapitalism**

Peter Taylor-Gooby: **The Double Crisis of the Welfare State and What We Can Do About It**

Jeffrey Meyers: **Remembering Iris Murdoch: Letter and Interviews**

Grace Ji-Sun Kim: **Colonialism,** Han, **and the Transformative Spirit**

Rodanthi Tzanelli: **Olympic Ceremonialism and the Performance of National Character: From London 2012 to Rio 2016**

Marvin L Astrada and Félix E. Martín: **Russia and Latin America: From Nation-State to Society of States**

Ramin Jahanbegloo: **Democracy in Iran**

DOI: 10.1057/9781137351937

palgrave▸pivot

Towards a Sustainable University: The Ca' Foscari Experience

Chiara Mio
Ca' Foscari University, Italy

palgrave
macmillan

DOI: 10.1057/9781137351937

First published 2013 by
PALGRAVE MACMILLAN

Palgrave Macmillan in the UK is an imprint of Macmillan Publishers Limited, registered in England, company number 785998, of Houndmills, Basingstoke, Hampshire RG21 6XS.

Palgrave Macmillan in the US is a division of St Martin's Press LLC, 175 Fifth Avenue, New York, NY 10010.

Palgrave Macmillan is the global academic imprint of the above companies and has companies and representatives throughout the world.

Palgrave® and Macmillan® are registered trademarks in the United States, the United Kingdom, Europe and other countries.

ISBN: 978-1-137-35194-4 EPUB
ISBN: 978-1-137-35193-7 PDF
ISBN: 978-1-137-35192-0 Hardback

A catalogue record for this book is available from the British Library.

A catalog record for this book is available from the Library of Congress.

www.palgrave.com/pivot

DOI: 10.1057/9781137351937

Contents

List of Tables

DOI: 10.1057/9781137351937

DOI: 10.1057/9781137351937

List of Figures

DOI: 10.1057/9781137351937

Acknowledgements

The author wishes to recognize the value of the input received from Corrado Clini (former Italian Minister of Environment, Land and Sea) and from Martina Hauser (Department of the Environment, Italy) and their staff: they contribute to the development of knowledge and experience around the world, both in sustainability and in carbon footprint management.

The author also wishes to recognize the enormous support received from colleagues working at Ca' Foscari – a long list of enthusiastic people supporting my commitment to sustainability: Prof. Carlo Carraro (Rector of Ca' Foscari), who has adopted, since the beginning, a strategic approach towards sustainability with outstanding performance; the team of Ca' Foscari University involved in sustainability (special thanks to Roberto Marin and Federica De Marco) – they are always open to new challenges, engaging other colleagues or stakeholders.

And special thanks to Magda Innocente for her contribution to the revision in polishing the language and finding appropriate translations.

DOI: 10.1057/9781137351937

1
Sustainability: The New Paradigm

Abstract: *Chapter 1 deals with the critical environmental and social themes that characterize the present scenario, generated and intensified by a development model characterized by a high growth rate, which has contributed to stressing the difference between conditions of 'wealth', to the advantage of the few, and conditions of 'poverty', which are progressively expanding, with alarming environmental repercussions in particular on the climate change front. Illustrated below are the main initiatives activated by countries, institutions and other organisms with the purpose of producing cultural change oriented towards sustainability and a review of the current development model, which takes shape in the endorsement of declarations and definitions of agreements with which commitments to sustainability are formalized. In this context, the universities have a very important role in the creation both of knowledge and of educational models that support this new sustainability paradigm.*

Mio, Chiara. *Towards a Sustainable University: The Ca' Foscari Experience*. Basingstoke: Palgrave Macmillan, 2013. DOI: 10.1057/9781137351937.

1.1 The scenario: the critical environmental and social themes

'Sustainable development is development that meets the needs of the present without compromising the ability of future generations to meet their own needs. It contains within it two key concepts:

- the concept of needs, in particular the essential needs of the world's poor, to which overriding priority should be given; and
- the idea of limitations imposed by the state of technology and social organization on the environment's ability to meet present and future needs.' (World Commission on Environment and Development, 1987)

This, quoted in the Brundtland Report, is the best-known definition of sustainable development, which should constitute the reference perspective in the affirmation of new production and consumption models, in the light of the world situation and of its criticalities at the economic, environmental and social level, destined to worsen in the absence of a different orientation of behaviour on the part of all the actors, from governments, institutions and organizations to individual citizens.

The principle of growth, which for decades has guided their actions, has led to a situation of over-production and over-consumption, with extremely negative consequences in environmental and social terms, in addition to economic ones.

At the environmental level, one of the issues with the highest criticality in the present context is climate change. The climate has always undergone changes due to natural causes, including solar radiation, volcanic eruptions and natural fluctuations of the climatic system in itself, but such phenomena can explain the increase of global warming only in part. A significant cause of global warming is represented by human activities, which have contributed significantly to generating a greater concentration of the greenhouse gases (GHG), such as carbon dioxide, methane and nitrous oxide, which intensify the natural greenhouse effect causing the warming of the planet. Among the consequences of climate change, in addition to global warming, is the raising of sea levels due to the melting of the polar icecaps, and an increase in floods and hurricanes, with repercussions on human life.

It is estimated that, in the absence of strong intervention on the matter, the total costs and the consequences stemming from non-sustainable

DOI: 10.1057/9781137351937

behaviour will amount to losses equal to 5% of global per capita consumption an estimated damage that will increase to 20% considering the broader range of risks and impacts, while the cost of the reduction of the ex ante emissions could be limited to only 1% of global annual GDP (Stern, 2006). The economic commitment necessary to rectify a phenomenon so devastating would therefore prove to be definitely less than the damages and the loss associated with it. The estimated increase in per capita GDP at a global level between 2001 and 2030 could be on average equal to 2.23% yearly (Maddison, 2005), but the fear is that this will be compromised by the shortage of energy sources and by the economic implications associated with global climate change, connected to the increase in emissions of carbon dioxide (CO_2) and of other GHGs, for example, in sectors such as agriculture, tourism and transportation. It appears extremely urgent for the various countries to reconcile the hoped-for economic recovery with a strong commitment towards energy efficiency, moreover favouring renewable sources over fossil fuels (Musu, 2009).

The present development model, like no other in the past, has been characterized by high growth rates that have contextually and dichotomously affected both the levels of wealth and the levels of poverty. Just over 14% of the earth's population accounts for almost 80% of the world consumption, entailing the production of over 50% of the total carbon dioxide and three times as much waste as the poorer 86% (Gesualdi, 2005). Therefore, the acceptance and integration of ethical aspects – in addition to the economic ones that have guided the decision-making process up to now – appear to be unavoidable. Also the public pressure on governments and decision makers is inspired by environmental and social issues.

A cultural change oriented towards sustainability in the relationships between man and the environment proves to be crucial (Schwarz J., Beloff B. and Beaver E., 2002), with a re-examination of the current development model, up to now supported by interpretative criteria of the cultural reality that are merely economic or quantitative. It appears necessary to drastically reduce the negative effects of economic growth, even going down a 'decrease strategy', concentrated on moderation, on the consideration of the existing limits and on the '8Rs' (recycle, reuse, reduce, redistribute, relocate, restructure, re-contextualize, re-evaluate), to attempt to respond to the serious current emergencies in environmental, social and economic fields (Latouche, 2007).

DOI: 10.1057/9781137351937

It is already obvious that developing a new way of living, producing and consuming, rethinking the relationships between men, and between men and nature, aiming to reconcile or at least balance the diverging interests of all the actors and stakeholders, is a fundamental condition for obtaining socially and ecologically responsible relationships and ties.

In the following chapter the main commitments and declarations endorsed by countries and institutions for the spreading and adoption of behaviour coherent with the sustainability paradigm are presented.

1.2 The commitments of countries and institutions

'The long-term maintenance of systems according to environmental, economic and social considerations' (Crane and Matten, 2004) requires a synergetic action concerted among the various countries, as well as among the various institutions, to support the process of cultural and educational transformation towards sustainable development.

Since the 1970s, several initiatives have followed one another, which have seen countries, institutions and other bodies involved with the endorsement of declarations and the definition of agreements formalizing precise commitments towards sustainability. At the institutional level, particularly significant is the role that universities have and are recognized to have (the subject of the present work) as fundamental actors in innovating knowledge, work methods and behaviour models, for effective development action and the promotion of a sustainable development culture.

Awareness of the educational role of the universities on the topic of sustainability is growing and to testify to it is the lively stream of international statements, the most important of which will be illustrated later on.

Becoming aware of the close link between development and the environment is the goal of a long path that started at the United Nations Conference on the Human Environment, held in Stockholm in 1972. The great changes which in that era disrupted the history of the world built a particular sensibility towards social and environmental themes. In the course of this meeting, important recommendations were formalized. Even if not mandatory, they aimed to inspire and guide people towards conservation and improvement of the context of reference, with a growing interest in the role of higher education in fostering a sustainable future.

DOI: 10.1057/9781137351937

An important stage along the path was the UNESCO Intergovernmental Conference on Environmental Education of 1977, held in Tbilisi in Georgia, organized by the United Nations Educational, Scientific and Cultural Organization (UNESCO) in collaboration with the United Nations Environmental Programme (UNEP). On this occasion an appeal was launched to the member states to introduce a series of measures with the aim of fostering and intensifying reflection, research and innovation concerning environmental education, partly through teaching and research and training courses.

In 1988, in Bologna, Italy, the Magna Charta Universitatum was signed by the representatives of the major European countries, a document that outlines the ideal characteristics of the university institution and identifyies as the purpose of such institutes the elaboration and preservation of knowledge and its transmission to subsequent generations, with the intention of broadening the limits of the expression of human potential (Rebora, 2007, p. 5).

The official start of the university 'sustainability movement' is marked by the Talloires Declaration of 1990. 'This is the first official statement made by university administrators of a commitment to environmental sustainability in higher education. The Talloires Declaration is a ten-point action plan for incorporating sustainability and environmental literacy in teaching, research, operations and outreach at colleges and universities. It has been signed by over 350 university presidents and chancellors in over 40 countries' (ULSF, 2013).

The ten points of the declaration were:

1. Increase awareness of environmentally sustainable development.
2. Create an institutional culture of sustainability.
3. Educate for environmentally responsible citizenship.
4. Foster environmental literacy for all.
5. Practice institutional ecology.
6. Involve all stakeholders.
7. Collaborate for interdisciplinary approaches.
8. Enhance capacity of primary and secondary schools.
9. Broaden service and outreach nationally and internationally.
10. Maintain the movement.

The main limitation of the Talloires Declaration is the almost exclusive consideration of the environmental dimension of sustainability, neglecting the social dimension and the economic dimension.

DOI: 10.1057/9781137351937

The 1991 Halifax Declaration is considered to be the first declaration that endorses the commitment of universities to sustainability development, intended in its tridimensional meaning: social, environmental and economic. With this document the importance of university actions in the spreading and implementation of sustainable development logics is confirmed. At the Halifax Conference, which was attended by presidents of universities from Brazil, Canada, Indonesia, Zimbabwe and elsewhere, as well as senior representatives of the International Association of Universities, the United Nations University and the Association of Universities and Colleges of Canada, the universities invited each other to commit themselves to the teaching of the principles and practices of sustainable development and of environmental literacy through a clear policy, apolitical and without compromises, to spread understanding of the environmental dangers and of the social inequalities that threaten the planet, pushing the new generations to better use of its meagre resources and to greater co-operation for a sustainable future.

The subsequent stage was in the following year, 1992, in Rio, with the United Nations Conference on Environment and Development (UNCED), the first world conference of the heads of states on the environment, an unprecedented event in terms of impact on policy choices and subsequent development, as well as in media terms. The main official documents it produced were the Rio Declaration on Environment and Development, and Agenda 21, a plan of action on sustainable development, to be carried out on a global scale by governments and by administrations of every kind, with the broadest possible involvement of stakeholders. In chapter 36 of Agenda 21 ('Education, training and public awareness'), it is declared that education has the essential role of fostering sustainable development and of improving the capability of people to relate with issues tied to development and to the environment (Calder and Clugston, 2003; Jones et al., 2010).

Also in 1992 the United Nations Framework Convention on Climate Change (UNFCCC) was approved, which intended to stabilize the concentrations of greenhouse gases in the atmosphere at a level that would prevent human activities from interfering negatively with the climate. The Convention iterates the "common but differentiated responsibilities" principle among industrialized and developing countries, establishing that the former are the first have to adopt concrete initiatives against climate change and its consequences, since they are primary people responsible for the present concentration of the greenhouse gases and

DOI: 10.1057/9781137351937

current holders of the means, both financial and technological, to reduce emissions. The signatories of the broad convention were obliged to establish national programmes for the reduction of emissions and to present regular reports regarding the implementation of the same.

In 1993, at the conclusion of the Association of Commonwealth Universities' Fifteenth Quinquennial Conference, held at the University of Wales, over 400 participants from 47 countries approved the Swansea Declaration, supporting the co-operation of all actors towards a common objective of sustainable development.

In the same year, with the Kyoto Declaration, promoted by the International Association of Universities (IAU) and adopted by 90 universities, the moral obligation of universities to carry out a reform was underlined, recommending specific plans of action at an institutional level with a view to sustainable development. It is one of the most important documents regarding the roles of the universities in achieving sustainability and sets out precise commitments and duties of these institutions, outlining for the first time in a formal way an ethical and social perspective of sustainability and identifying the appropriate lines of action.

Another important step was the CRE-Copernicus Charter of 1994, with which the signatory universities (312 universities of 37 countries) collectively undertook to foster better comprehension of the interaction between man and the environment, in addition to collaborating on commom environmental themes.

The Thessaloniki Declaration 'Environment and Society: education and public awareness for viability' was approved in 1997 in Salonicco, Greece, at the conclusion of the International Conference organized by UNESCO. In the declaration, signed by 90 countries, governments worldwide were recommended to develop suitable educational tools to ensure a sustainable future, allocating financial resources to the education and awareness of citizens, orienting all the disciplines towards sustainable development, through a holistic and interdisciplinary approach.

1997 was also the year of the Kyoto Protocol, an international agreement connected to the UNFCCC that committed the adhering countries to internationally binding emission reduction targets. The European Community (EC) proved to be more committed to the Protocol than the United States, which, even though being responsible for more than 35% of the global emissions, did not even sign the agreement. The latter instead adopted different approaches and systems to cope with the

DOI: 10.1057/9781137351937

challenge of climate change. The EC, for example, introduced a 'cap-and-trade' system of reduction and exchange of carbon emissions, which is a system based on the regulation or limitation of emissions (cap) and a contemporaneous market (trade) in emission quotas allotted to organizations or units governed by the system (factories, power stations and other installations identified by the government of each country). These units were obliged to control their emissions within identified limits or, alternatively, if considered more expedient, to purchase through certain institutions corresponding 'emission credits' that allowed them to go beyond the established threshold. Other high-impact economies, like Japan and China, also set emission reduction targets. In the United Kingdom, a Carbon Tax has been in force since 2001, applied to business consumption and those of public operations in the gas, coal, electricity and petrol sectors, to incentivize these organizations to improve their energy efficiency and to reduce emissions.

In the wake of the Kyoto Protocol various initiatives have been made, including the 2002 Carbon Disclosure Project (CDP), launched by a non-profit organization (Rockefeller Philanthropy Advisors) and supported by 400 institutional investors. The project has the goal of expediting solutions to climate change by coherently orienting investment decisions through the evaluation of the climate-change strategies adopted by the major global companies. The aim is to affect investors' decisions and then capitalize on the stock exchange by considering the environmental performance evaluation of the company. On behalf of its investors the CDP monitors yearly the energy policies and performance of businesses (which adhere on a voluntary basis) and their capabilities to manage the economic opportunities and the climate risks, with the intention of encouraging the organizations to measure, protect and make public their greenhouse gas emissions.

In 1998 the United Nations Commission on Sustainable Development and UNESCO adopted the World Declaration on Higher Education for the 21st Century: Vision and Action, where the centrality of higher education was affirmed, 'as well as an increased awareness of its vital importance for sociocultural and economic development, and for building the future, for which the younger generations will need to be equipped with new skills, knowledge and ideals', pointing out that 'the core missions and values of higher education, in particular the mission to contribute to the sustainable development and improvement of society as a whole, should be preserved, reinforced and further expanded' (United Nations Educational, Scientific and Cultural Organization, 1998).

DOI: 10.1057/9781137351937

Another important stage was marked in 2001 with the Lüneburg Declaration, on the occasion of the International Copernicus Conference 'Higher Education for Sustainability – Towards the World Summit on Sustainable Development (Rio+10)' adopted by the Global Higher Education for Sustainability Partnership (GHESP), a body recognized by UNESCO that incorporates the Association of Unviersity Leaders for a Sustainable Future (ULSF) and International Association of Universities (IAU), connecting to the Copernicus-Campus. The goal of the declaration was that of further mobilizing the international university network on the sustainability theme.

2002 was the year of the World Summit on Sustainable Development in Johannesburg, which tens of thousands of people attended, including heads of state and of government, national delegates and leaders of non-governmental organizations (NGOs), businesses and other bodies, whose aim was 'to focus the world's attention and direct action toward meeting difficult challenges, including improving people's lives and conserving our natural resources in a world that is growing in population, with ever-increasing demands for food, water, shelter, sanitation, energy, health services and economic security' (United Nations, 2002b).

Also in 2002 the Ubuntu Declaration 'Education and Science and Technology for Sustainable Development' was signed, on behalf of the United Nations University (UNU), UNESCO, the IAU, Third World Academy of Sciences, African Academy of Science, Science Council of Asia, International Council for Science, World Federation of Engineering Organizations, Copernicus-Campus, Global Higher Education for Sustainability Partnership (GHESP) and ULSF. The declaration pointed out the need for the integration of the themes of sustainable development in the course of education, starting from primary school, as a fundamental basis for the pursuit of sustainability objectives.

In the same year, in the course of the 57th General Assembly of the United Nations, the period from January 2005 to December 2014 was proclaimed the Decade of Education for Sustainable Development (DESD), to confirm the centrality of education in the process towards the paradigm of sustainable development, meaning by Education for Sustainable Development (ESD) a process oriented towards human development based on three pillars proposed by the United Nation Development Programme (UNDP), namely economic growth, social development and protection of the environment. The basic vision of ESD is that of a world where everyone can benefit from quality education and

DOI: 10.1057/9781137351937

learn values, behaviour and a lifestyle coherent with sustainable development and the improvement of society.

In Rio de Janeiro in 2012, 20 years after the Earth Summit of 1992, the United Nations Conference on Sustainable Development (UNCSD) Earth Summit 'Rio+20' took place, with thousands of participants, including world leaders, heads of government, the entrepreneurial world, NGOs and other bodies and institutions. The main themes of the conference were twofold: the green economy in the context of poverty eradication and sustainable development, and the institutional framework for sustainable development. These two themes were yoked to seven overriding 'scopes of action': decent jobs, energy, sustainable cities, food security and sustainable agriculture, water, oceans and disaster readiness, which led to the formalization, at the end of the summit, of the Outcome Document 'The Future We Want' (United Nations, 2012), delivered to more than 100 heads of state and government.

This path of commitment undertaking, outlined in chronological order, points out a growing interest and commitment on the part of countries, institutions and the public in the sustainable development theme. In the last 40 years several initiatives and projects of various types have followed one another, some wide-ranging, which have involved countries worldwide, others more targeted, including those concentrated on the role, considered fundamental, of education in the pursuit of the objectives of sustainable development, to trigger the transition of socio-economic reality towards a more sustainable lifestyle. Particularly active in these commitment declarations were university institutions, who focused their attention on sustainability themes that, as we have seen, developed and spread coherently with the evolutionary path embarked on by countries and other institutions. This testifies to an awareness on the part of the different actors of the importance of embarking on a path to sustainability in a synergetic and concerted manner, in which the definition of the underlying policies on the part of the supranational bodies and countries constitutes the framework into which the commitment and the action of the individual institutions must fit. Universities in particular must pass on and spread the new paradigm, for an effective comprehension of the sustainability logics on the part of the community, sensitizing students and fostering reflection, as well as activating themselves directly. Because of this it is fundamental to trigger a change in the university institution, through the adoption of strategies marked by sustainability, which coherently address daily behaviour at all levels

DOI: 10.1057/9781137351937

of the organization and directed towards stakeholders. The educational action carried out by universities is one of the central driving factors in the spreading of the sustainability perspective, through the training of citizens to be aware of their role in the community and to adopt sustainable behaviour, citizens who are suitably informed and motivated, as well as ready, for the change that awaits them and that will see them as protagonists.

These themes will be the object of closer examination in the following chapters.

DOI: 10.1057/9781137351937

2
New Responsibilities for Universities

Abstract: *Chapter 2 deals with the sustainability theme in universities and with the main factors that have contributed to the principles and characteristics of the sustainable university (including Principles for Responsible Management Education) and the central elements of the two approaches to sustainability strategy (technical-technological and strategic-organizational). Consequently the strategic-organizational approach is investigated by analysing: governance systems and strategy, along with stakeholder engagement; performance measurement and evaluation systems, along with key performance indicators (KPI) of sustainability and the centrality of the outcome indicators in the set of KPI; and reporting systems, along with sustainability reports and the new approach of integrated reporting. Therefore, two themes are dealt with, which point out the commitment of universities: the carbon management process (along with the carbon footprint) and green universities, along with sustainability networks and rankings.*

Mio, Chiara. *Towards a Sustainable University: The Ca' Foscari Experience*. Basingstoke: Palgrave Macmillan, 2013. DOI: 10.1057/9781137351937.

 DOI: 10.1057/9781137351937

2.1 Sustainability in universities

Companies are considered largely responsible for the increase in pollution and for the affirmation of the social model of production and consumption that is currently widespread, characterized by high-intensity use of resources, mostly not renewable. However, the role that other types of organization play in pollution is undeniable and they can also play a part in the spreading and application of sustainability concepts.

Of particular significance in the last few decades has been the commitment of companies to the doctrine of proposing contributions in which the importance of the sustainability paradigm in universities is pointed out with vigour, proceeding substantially in parallel with the various initiatives proposed at the international level, which were illustrated in Chapter 1.

Below, some of the main scientific contributions proposed are presented and the central characteristics of two paradigmatic typologies of approaches to sustainability are defined, pointing out the objective model towards which universities must aspire.

2.1.1 The importance of sustainability in universities: Some contributions

Besides consuming resources in the execution of their activities and therefore directly generating environmental and social impacts, university institutions, as pointed out in the Brundtland Report of 1987 (which also refers to educational structures, from kindergarten to high level education), hold a central role in spreading the sustainability concept in the community, since they have the duty to train and to educate future generations to adopt this paradigm – those that will become the protagonists of the economic growth of tomorrow (MacVaugh and Norton, 2012).

University institutions therefore have a social responsibility that is intrinsic to their purpose and their actions as regards evaluation that transcends time and the boundaries that are familiar to entrepreneurial organizations. In universities, as in general in the civil service, the mission itself is strongly infused with concepts tied to sustainability. For university institutions, therefore, their raison d'être requires an overall orientation that is coherent in this sense, from strategic approach to governance system to operational mechanism, that is the performance

DOI: 10.1057/9781137351937

measurement system, the evaluation system and the reporting system. The missions of universities, however necessarily very different among the various realities, are associated by sustainability themes, even with different descriptions and underlying connotations. If traditional aspects emerge in the first place, as for example the core issues represented by didactics and by research, even the strong references to the external context, to the territory where the university operates, are widespread. For example, with commitment to the spreading and promotion of the results of the educational processes and of scientific research in civil society, to the subjects that are part of the community, at both the national and international level. In the mission there is a tendency to affirm explicity the university's role in the context of reference, with respect to a plurality of stakeholders – not only the students, the professors and the technical-administrative staff, but also the entire community and the various actors that operate in it, from the entrepreneurial fabric to the civil service. In their missions, universities therefore recognize and assert the essence of the sustainability concept, recognizing a basic role in this perspective. This characterization in terms of sustainability, intrinsic in the university institution's raison d'être, is then confirmed in the axes of strategic development of this type of organization. This is because a sustainability strategy cannot be expressed in mere terms of compliance, but rather it is accompanied and supported by a commitment at a strategic level, in order to stimulate a process of transformation that permeates the value and cultural fabric of the organization (Mio and Borgato, 2012).

On the other hand, the explicit acceptance of sustainability principles, with reference to the attention directed at the internal and external stakeholders, in the university's mission, in its values and in its strategic plan, can constitute a fundamental lever, not only at the motivational level of the individuals involved, but also in terms of the capability to attract students and financial support (Marshall et al., 2010).

Considering the situations of extreme criticality, from environmental deterioration and economic crisis and uncertainty to social inequalities, educational institutions, and in particular universities, must educate and train students to face, manage and find a remedy for, as far as possible, these criticalities by defining and applying new theoretical paradigms, improving man's quality of life and engaging the whole community in the path towards a more sustainable future (Gadsby and Bullivant, 2010; Jones et al., 2010). Nelson Mandela refers to education as 'the most

DOI: 10.1057/9781137351937

powerful weapon which you can use to change the world' (Rio+20, para. 48).

The centrality of the academic world in the spreading of the principles of sustainable development was further confirmed in the course of the already cited Conference of Rio in 2012: 'We recognize the important contribution of the scientific and technological community to sustainable development. We are committed to working with and fostering collaboration among the academic, scientific and technological community, in particular in developing countries, to close the technological gap between developing and developed countries and strengthen the science–policy interface as well as to foster international research collaboration on sustainable development' (United Nations, 2012, p. 9).

The role of the university as regards sustainability is vast. It goes beyond the consideration and promotion of this paradigm in the course of curricular studies, since it must contaminate also the activities of research, stimulating individual and collective reflection and behaviour, increase the intellectual, emotional and political commitment of students towards sustainability (Jones et al., 2010, p. 164) and stimulate processes of collaboration among students, fostering a sustainable education.

The universities therefore have an extremely important and demanding responsibility, which is that of increasing the necessary awareness, knowledge, technologies and tools to create a sustainable future from the environmental point of view (Calder and Clugston, 2003), naturally in addition to sociality, the dimension that is more directly referable to the university institution in the first instance.

The over 400 universities in 50 countries that subscribed to Talloires Declaration (mentioned in Chapter 1) defined a voluntary plan of action, consisting of ten fundamental points, to build a sustainable university. In these ten points, therefore, the constituent elements of a sustainable university are recognized (University Leaders for Sustainable Future, 1990).

Other fundamental conceptual elements for the acceptance of sustainability logics in universities are found in the contribution provided in 2002, during the World Summit on Sustainable Development in Johannesburg, when the objectives and steps to be taken to bring high-level education towards sustainability were precisely defined (United Nations, 2002a):

- To integrate information and communication technologies inside the curricula, optional courses on sustainability are necessary,

DOI: 10.1057/9781137351937

but the education to sustainability must be incorporated into each curriculum in all the disciplines and should go beyond the teachings in the classroom to support the university in the actions focused on sustainability (Shriberg, 2002);
- To foster in the developed countries access to study programmes at affordable prices for students and lecturers coming from developing countries in order to incentivize the exchange of knowledge and experience;
- To continue to implement the work programme with the Commitee on the sustainable development on the education for sustainable development (ESD).

'A more sustainable university is a higher education institution, as a whole or as a part, that addresses, involves and promotes, on a regional or a global level, the elimination and/or minimization of environmental, economic, societal, and negative health effects in the use of their resources in order to fulfill its main functions of teaching, research, outreach & partnership and stewardship, among others, as a way to help society shift to a more sustainable way of living.' (Velazquez et al., 2006).

The role of the university institution as regards sustainability is central and widespread, since the global, economic, social and environmental crisis is first a crisis of values, ideas, perspectives and knowledge (Cortese, 2003). In this sense the universities are called upon to spread sustainability principles, not only through teaching and research, the core activities of these institutions, but also in every action, including relations with various stakeholders. It is a matter of creating and strengthening awareness, knowledge, capabilities and values to succeed in creating a fair and sustainable system, exploiting the academic freedom and the numerous and heterogeneous capabilities available that permit the development of new ideas and commit people to the search for solutions that favour sustainable lifestyles. The university must operate as an organization that intergrates social and environmental sustainability and manages interdependency with local, national and global communities: the relationships between teaching activities and research on the one hand and the local community on the other are undeniable, and universities, by reason of the intrinsic social responsibility that characterizes them and that is expressed in their mission, have a sort of 'moral obligation' to those communities (Locatelli and Schena, 2011).

DOI: 10.1057/9781137351937

The university system must coherently structure all its constituent elements to succeed in spreading the principles of sustainable development, since the various stakeholders, beginning with the students, the first recipients of university action, are currently exposed to numerous inputs and stimuli from various communication tools that are not necessarily coherent with a sustainable lifestyle, from the social, economic and environmental point of view. In particular, the university must succeed in (Cortese, 2003):

a) Making the deep connection between curricular studies and research stand out, to be planned and implemented in a synergic perspective, strengthening the sustainability contents;
b) Implementing measures aimed at reducing the university's negative social and environmental impacts, spreading the results achieved;
c) Producing positive effects for the community, in order to improve the quality of life, the level of economic security and environmental sustainability of the activities realized in the territory to which it belongs.

Managing to understand the functioning modalities of the environment and the limits that man must live within, without undermining the survival of the planet, is one of the objectives of education in the 21st century. This requires the development of specific knowledge and expertise, which must accompany and integrate with social and economic aspects, to create a fair and sustainable society. The university environment represents a sort of 'community', whose modalities of acting daily can constitute an essential vehicle in the transmission of sustainable life-styles to the various subjects, contributing to the reinforcement of a desire for values, positive behaviour oriented towards respect for the environment and for society, developing a sense of collaboration in the whole community.

A greater involvement of students in the learning experience is a driving factor in their personal growth, incentivizing their sense of belonging and the attention to the local community and ecosystem. As will be seen, investing in the active role of students, stimulated through an effective engagement process that allows them to apply what has been transferred through the teaching activity, desirably strengthened by the possibility to devote themselves to specific research on sustainability themes, can facilitate the formation of their own approach towards social responsibility.

DOI: 10.1057/9781137351937

This can also allow them to look with a critical eye at obsolete models (MacVaugh and Norton, 2012), which do not consider aspects (that are not possible to be ignored, at this stage), such as the limited nature of natural resources or the irreversibility of environmental deterioration (Marshall et al., 2010).

There are numerous positive repercussions coming from the proactive acceptance of the sustainability logics by universities, a crucial contribution to the creation of a sustainable society, including (Cortese, 2003):

- Improvement in instruction for all the stakeholders, internal and external to the university – instruction intended in the wide sense, also in terms of transference of principles for a sustainable life-style;
- Training of competent students, not only as regards the professional field, but also in their roles as responsible citizens;
- Higher level of recognition of the role of the university by external subjects;
- Greater power to attract students, capabilities and assets;
- Reduction of costs, not only in economic terms, but also from the environmental and social points of view;
- Activation of greater co-operation among the various subjects, for a shared contribution to the creation of a sustainable society, in the awareness that sustainability requires the contribution of all stakeholders.

A university institution commited to sustainability should, among other things, train and educate students so that they can understand the causes of the present environmental deterioration, the social inequalities and the injustice that often charaterize the community, motivating them to act with actions aimed at sustainability, making a contribution to the creation of a more just society, more respectful of the ecosystem (Clugston and Calder, 1999). Among the universities that declare a commitment to sustainability, the Association of University Leaders for a Sustainable Future has noticed some common traits, in terms of the typology of the actions implemented, where the commitments taken on become concrete. These include:

- Explicit acceptance of sustainability in the mission, in the values and commitments of the university and the attention directed at the internal and external stakeholders (Marshall et al., 2010);
- Sustainability that finds expression in the definition of an educational offer with specific contents, as well as in the realization

DOI: 10.1057/9781137351937

of seminars, conferences and in general in the activities organized by the university;
- The possibility for students to actively participate in university activities, the result of a reflection that redefines the role of the university institution, a considerable change as regards the traditional accademic model, which gives students a 'passive' role, confined to the teaching activity;
- Adoption of selection processes and assigning of job roles and career advancement also based on sustainability principles;
- Translation in specific actions of the sustainability policies contained in the mission into values and into a declaration of commitment, through, for example, measuring and monitoring the university's carbon footprint (which will be discussed later on);
- An offer of services of support and help to academic life and in the university campuses that gives the declared policies concreteness, including the orientation of new students, the establishment of structures devoted to sustainability, regular evaluations of the environmental implications of university activities, and the organization of public events, open to students, to members of staff and to the community for the spreading of the sustainability themes;
- Commitment in the formation of partnerships, both at the national level and at the international level, for the spreading of sustainability principles.

In particular, an approach of university openness towards the community, which recognizes their close connection, an approach that emerges and is enhanced in the creation of partnerships, as well as in the consideration of expectations and needs expressed and concealed in the context of reference, is a driving factor for a more widespread legitimation of the university's work, as regards the adoption of more 'prescriptive' approaches that risk not being in alignment with the reality and therefore less effective (Savan and Sider, 2003).

Among the different faculties, those of an economic nature have a particuarly important role in the process of changing students' way of thinking and acting, towards a model of sustainable development (Marshall et al., 2010). Specifically there are three courses to pursue in a sustainability perspective:

- Accepting this new paradigm, which translates into a new 'way of thinking', s substantial change in approach and in the way of

DOI: 10.1057/9781137351937

behaving that charaterizes many of the faculties belonging to the economic area. These must provide knowledge and tools to point out and define solutions to interdisciplinary questions, that evaluate the interconnections among the various economic scopes and the non-economic ones, focusing on the real and complex problems. The fact of fostering collaboration among professors, making the connection between the various disciplines closer, generates advantages for all stakeholders, as well as for the university itself;

- Carrying out scientific research on the sustainability theme, to trigger synergies, getting past departmentalized economic visions, which are damaging in a sustainability perspective;
- Enhancing the institution's coporate and human capital. Projecting the traditional education in economic subjects towards a broader connotation of improvement of the corporate capital increases the possibility of recognizing and appreciating interdisciplinary research projects, to the benefit of all stakeholders, as well as teachings also targeted at transferring problem-solving concepts and approaches in complex situations (MacVaugh and Norton, 2012). The attention paid to human capital increases, as does the sense of trust in the faculty and between the faculty and the students and staff. Thus, students are encouraged to actively participate in university life, legitimating and supporting the university in the achievement of the sustainability objectives. In support of the changes concentrated on the students, the university staff – in particular those that hold a managerial role – must prove to be favourable to and inclined towards the requested transformation, in effect leading their collaborators, trying to balance their needs and expectations in the decisional processes, and acting to maximize the value created.

A fundamental contribution that further consolidated the importance of sustainability for the academic world was the initiative Principles for Responsible Management Education, which led to the formulation of a framework for universities, in the perspective of orienting these institutions towards the sustainability perspective through the adoption of a gradual but systematic approach: 'The mission of the Principles for Responsible Management Education (PRME) initiative is to inspire and champion responsible management education, research and thought leadership globally. The PRME are inspired by internationally accepted

DOI: 10.1057/9781137351937

values such as the principles of the United Nations Global Compact. They seek to establish a process of continuous improvement among institutions of management education in order to develop a new generation of business leaders capable of managing the complex challenges faced by business and society in the 21st century. In the current academic environment, corporate responsibility and sustainability have entered but not yet become embedded in the mainstream of business-related education. The PRME are therefore a timely global call for business schools and universities worldwide to gradually adapt their curricula, research, teaching methodologies and institutional strategies to the new business challenges and opportunities' (Principles for Responsible Management Education, 2013).

In the six Principles the central elements that distinguish the 'strategic organizational' approach to sustainability are grasped, and these form the reference of the present work and will be the subject of closer examination in the following paragraphs.

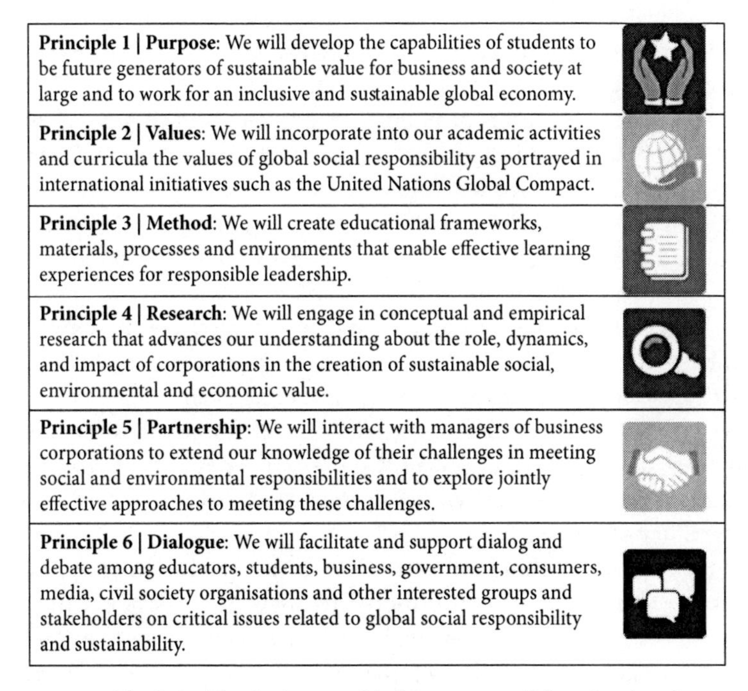

Principle 1 | Purpose: We will develop the capabilities of students to be future generators of sustainable value for business and society at large and to work for an inclusive and sustainable global economy.

Principle 2 | Values: We will incorporate into our academic activities and curricula the values of global social responsibility as portrayed in international initiatives such as the United Nations Global Compact.

Principle 3 | Method: We will create educational frameworks, materials, processes and environments that enable effective learning experiences for responsible leadership.

Principle 4 | Research: We will engage in conceptual and empirical research that advances our understanding about the role, dynamics, and impact of corporations in the creation of sustainable social, environmental and economic value.

Principle 5 | Partnership: We will interact with managers of business corporations to extend our knowledge of their challenges in meeting social and environmental responsibilities and to explore jointly effective approaches to meeting these challenges.

Principle 6 | Dialogue: We will facilitate and support dialog and debate among educators, students, business, government, consumers, media, civil society organisations and other interested groups and stakeholders on critical issues related to global social responsibility and sustainability.

FIGURE 2.1 *The Principles for Responsible Management Education (2013)*

DOI: 10.1057/9781137351937

2.1.2 Approaches to sustainability

Numerous contributions have established and recognized, at a scientific doctrinal level, the importance of sustainability for university institutions. All these contributions aim at defining the characteristics where an attention towards sustainability can be recognized, even if there are differences among the procedures adopted by universities on this theme.

A variety of approaches emerges, which can be placed within a continuum of solutions, whose extremes are represented by two possible configurations:

- Technical-technological approach;
- Strategic-organizational approach.

The first extreme identifies an approach towards sustainability that is made concrete in single projects and initiatives, with one-off value, in the absence of a unitary and widespread perspective, strategically oriented and consistently supported at the managerial and organizational level. The definition of this approach is tied to the fact that the projects activated generally fit within a scope of a 'technical-technological' nature mainly oriented towards the environmental dimension. For example, they usually become a concrete reality in the adoption of clean technologies, as in the installation of photovoltaic panels and in general green building initiatives, as in the case of heat insulation, in the replacement of traditional systems for heating, air-conditioning and lighting by systems with a smaller environmental impact, for greater energy efficiency of the buildings, with positive impacts also on the emissions generated (therefore a projection towards the carbon management approach, nevertheless without its systematic and widespread nature, as will be seen later on), or in the handling of waste, with the implementation of a system of selective collection, or more initiatives of sustainable mobility, or systems aimed at the sustainable use of water resources. In the case of initiatives aimed at greater sociality, these can concern the involvement of students in voluntary activities in favour of organizations that work in the territory of reference.

At the opposite extreme there is the 'strategic-organizational' approach, which represents the objective model to strive towards for a full and coherent acceptance of the sustainability logics. Such an approach, which will be the subject of closer examination in the next chapter, entails, for the university, a demanding intervention of a comprehensive nature,

DOI: 10.1057/9781137351937

widespread in that it proceeds from a unitary perspective and goes on to involve the aspects of a strategic, managerial and organizational nature, with deep repercussions, and spreads to all the fronts where the university's actions engage.

The adoption of the strategic organizational approach, (as in the continuous improvement approach) that distinguishes the paradigm of sustainability, can never be considered fully accomplished; therefore, it represents an objective model to strive for but is never completely achieved.

According to this approach, sustainability must first emerge in the elements where the university's long-term path is formalized, that is the mission, the value system, to then be described coherently in the objectives and in the lines of strategic policy and therefore in the short-term objectives, infiltrating the governance system, which must 'widen' to accept and understand the expectations of the various categories of stakeholder, going beyond the limited boundaries where traditionally the institutional structure is defined. As a result, the organizational structure and the various operational mechanisms must be modelled on the dimensions of sustainability, from the system of performance measurement, to the evaluation system, to the reporting system, to the management system of human resources, and so on. If, as has been said, sustainability is intrinsic to the very mission of the university, it cannot be that its translation into actions can emerge automatically as a result of the sensitivity and the initiative of the individual actors or groups that are a part of it. Instead, it is necessary to plan an intervention according to a unitary perspective and targeted in a manner shared with the stakeholders, with the strong imprimatur of the leaders: the pursuit of the organizational-strategic approach is therefore a path the requires time to become established and to generate significant results but is the only rewarding one at the level of value creation, in its broadest meaning.

With this approach, the fundamental principle of 'do as you say' is expressed, which is unavoidable to give credibility and substance to statements expressed by the university, formalized and transmitted to stakeholders through various tools, so that the 'talk' does not end in one-off statements and reporting, perhaps on single themes, towards which at a certain moment public opinion has a higher sensitivity, but fits in a framework of coherent sustainability, whose axes are identified and transferred into the mission, into the values and into the strategic policies. The activation of actions aimed at translating and making the

DOI: 10.1057/9781137351937

declared orientation towards sustainability visible in deeds is fundamental so that this is considered believable and does not remain in the realm of mere statements of intent, creating trust among stakeholders and activating that positive loop, source of precious synergies, that comes from the widespread adoption of sustainable behaviour by all stakeholders in the territory of reference, to that extent sensitized to and 'educated' in such sense by the university.

The imaginary line that joins the two extreme approaches identified, 'technical-technological' and 'strategic-organizational', can identify a sort of path for the university towards sustainability. It is not rare, indeed, that in the initial phase a definition tends to prevail that is made concrete in single projects, for a gradual moving closer to the new paradigm – projects through which one begins, so to speak, to 'get to know' the sustainability principles (as happens in other organizations). If, however, the organization remains immobilized within this approach, that is if it continues to proceed by single initiatives, the positive repercussions that it will be able to generate in the context of reference will be limited, because they are misaligned with sustainability perspective, which require a systematic approach.

If therefore, in a first phase, a 'technical-technological' approach is justifiable, it is important that the university be aware that this does not mean to be sustainable. The university's 'sustainability agenda' must therefore provide for a series of evolutionary stages, that permit to project the university towards the adoption of the strategic-organizational approach, the only one that is really effective for the creation of lasting and sustainable value.

2.2 The strategic-organizational approach

The 'strategic-organizational' approach represents the objective towards which the university must project itself for a complete acceptance and effective implementation of the sustainability paradigm. It is substantially a matter of defining as a whole the reason for existence and the acting logics according to the principles of sustainable development. If, as revealed previously, sustainability by its nature falls within the university's mission, its concrete translation into actions requires the coherent orientation of all its components, proceeding from a unitary perspective. This means intervening in the various systems where the macro system

DOI: 10.1057/9781137351937

of the university can be incoherent; among these, particularly critical is intervention in the governance systems, strategic planning, measurement and evaluation of performance, and reporting. Below are presented the main aspects related to these systems, along with tools and actions that are connected to these, which must be activated to strengthen the effectiveness of the path towards sustainability.

2.2.1 Sustainability, governance and strategy

A first fundamental intervention for the acceptance of sustainability logics stresses the organization and its systems, which permit a definition of the university's long-term path, expressing the necessary wide-ranging perspective that must characterize it, as well as an openness beyond the university's institutional boundaries.

First of all, the acceptance of the sustainability principles and the broadening of the governance system entail repercussions at the organizational level. On the one hand, fundamental to ensure the success of the path towards sustainability is the strong imprimatur of the university's leaders, as well as their presence and participation overall in particularly significant moments – for example, in the presentation of important sustainability projects that involve the university. If the university is governed by a collective body, the role of the rector, which resembles that of the chief executive officer of businesses (Sammalisto and Arvidsson, 2005), is in any case central considering the growing independence that the universities have been acquiring in the last few years. The rector therefore has a central role in areas such as the definition of the mission and the activation of partnerships with stakeholders, as well as in decisions on the modalities of fund allocation (Trani and Holsworth, 2010).

On the other hand, so that the principles and the logics of the sustainable acting are effectively spread and interiorized at all the levels of the organization, it is necessary to provide for total governance of the mechanisms of the sustainability dynamics, by assigning precise responsibilities at a level below that of the governing bodies, but, all the same, leaders, usually with the establishment of dedicated organizational figures and bodies that have the task of co-ordinating the realization of the relevant activities of the university, protecting the relations with stakeholders. This can be made concrete with the nomination of a university 'head of sustainability', a figure who is part of the leadership team and who is usually chosen from among the accademic staff and

DOI: 10.1057/9781137351937

acts through the authorization of the rector. The university can also opt for a collective body, a specially established commitee or an external body. This figure (or body) represents the point of reference for internal and external stakeholders; it is the conerstone in the organization for the monitoring of the implementation of the sustainability policies and the spreading of projects on relevant themes. Where it is an individual, the figure can be duly supported by an organizational unit dedicated to sustainability, which is given tasks in processes and projects transversal to the organization, acting as a collector of information, for example, in the writing of sustainability reports. In terms of procedure, in some universities, bodies were also established, such as a student council or joint committees, which permitted the active involvement of certain categories of stakeholders.

On the level of the governance system, the acceptance of sustainability logics entails a 'broadening' of the system, including the various categories of stakeholder, allowing categories of subjects traditionally excluded from the main decisional circuits to intervene, according to differentiated modalities and tools, to protect their own rights and their own requirements, therefore activating a widened governance process (Mio, 2005).

According to what is still the best-known definition, Freeman's, 'a stakeholder is every individual identifiable that can influence or be influenced by the activity of the organization in terms of product, policies and work processes' (Hinna, 2002, p. 405, own translation). According to the Global Reporting Initiative, 'stakeholders are defined broadly as those groups or individuals: (a) that can reasonably be expected to be significantly affected by the organization's activities, products, and/or services; or (b) whose actions can reasonably be expected to affect the ability of the organization to successfully implement its strategies and achieve its objectives' (GRI, 2011, p.43).

The identification of stakeholders represents an element of criticality for the university, from the moment that it decides to activate an evolutionary path towards sustainability. When placed in the sustainability perspective, the process of mapping stakeholders, necessarily personalized and defined for the purpose by each university, is characterized by a high degree of complexity in space and time. This is because the sustainability concept, as has been mentioned, requires going beyond the boundaries of the organization (space), and facing the expectations of future generations (time). The recognition by the university of a subject

DOI: 10.1057/9781137351937

as a stakeholder and the degree of interdependency that is established with it, is affected by the values, the mission and the basic strategic orientation of the university, as well as by their concrete translation into decisions and actions.

The mapping of stakeholders is therefore an essential step in the pursuit of sustainability: it is not only a matter of identifying the categories of those having an interest at a formal level. It is also the introduction of a category of subjects in the stakeholder mapping implies the consequent consideration and acceptance of the relevant expectations, variously expressed and accepted by the university, being expressed in the policies, programmes and actions introduced. The same degree of detail chosen for the mapping, the splitting of broader categories into sub-categories, the perspectives and the same modality with which they were identified, do not have a merely descriptive value, but constitute a specification motivated by the university's will to grasp, to represent and to accept distinct expectations in one category, to which engagement tools and actions correspond.

The mapping of stakeholders is indeed a functional step towards the activation of intervention aimed at satisfying (as far as it possibly can) the requirements and the scope. The latter has numerous factors of complexity, since the stakeholders identified present requirements that are not only different, but frequently also conflicting, and for which the university must look for a balance, in the context of scarce resources at its disposal. The activation of a stakeholder engagement process is a fundamental, constituent element in accounting and is a necessity for stakeholders (Locatelli and Schena, 2011). The stakeholder engagement is defined as a process of 'construction of a permanent dialogue' (Hinna, 2002, p. 329), structured and continuous, constantly used by the organization, not only in the reporting phase (Hinna, 2002, p. 329), which makes use of different tools, (depending on the intensity and the frequency of the relations established with the various categories, but among which, for example, are: interviews, questionnaires and group meetings with opinion leaders identified within each category, and surveys on specific topics) (Hinna, 2002). Activating a stakeholder engagement process allows the university to achieve results on different fronts, for example, in terms of effectiveness of the understanding of the dynamics of where it's operating and the subjects which, directly and indirectly, are affected by the activities carried out; the possibility to use the inputs of stakeholders to improve processes and services supplied; the education of the stakeholders

DOI: 10.1057/9781137351937

towards sustainable development; spreading of responsible lifestyles; an increase of the trust between the organization and the stakeholders and consequently of legitimation by the same, as well as in terms of risk management (AccountAbility & United Nations Environment Programme Stakeholder Research Approach, 2005b; AccountAbility, 2011).

The stakeholder engagement can be variously structured, adopting different tools four fundamental types that entail different degrees of commitment from the university in terms of time, investment, risk and willingness to collaborate (AccountAbility & United Nations Environment Programme Stakeholder Research Approach, 2005a):

- Reporting, which consists of the transmission of information of various types to stakeholders, for example regarding the organization itself, the activities carried out, planned projects. To be effective it must be honest, transparent, timely and precise. Reporting tools are, for example, brochures and reports, newsletters, websites, conferences and press releases;
- Consultation with stakeholders and the collection of information and suggestions from stakeholders, whose results are used in the decision-making processes and in the planning of activities. Usually the stakeholders' perception of the organization and of the impacts of the activities introduced, and their reflections on current and emerging issues are the object of consultation. Tools of consultation are, for example, questionnaires, focus groups, online feedback and discussion forums;
- Dialogue, which implies an exchange of opinions and points of view between the organization and the stakeholders. Unlike consultation, dialogue is not carried out by the organization but it is two-way: on the one hand, the stakeholder has a greater power to influence the counterparty; on the other hand, the organization has the possibility to contextualize its choices and declarations. Dialogue, to be effective, presupposes an objective and unbiased vision and the will to put aside prejudice and to listen to opinions, even conflicting ones, and is greatly influenced by the cultural context and by the attitude of the subjects that communicate. Tools of dialogue are, for example, multi-stakeholder forums, meetings with leaders and advisory panels;
- Partnership, an agreement between the organization and one or more subjects that commit themselves to the pursuit of common

DOI: 10.1057/9781137351937

objectives, placing expertise and resources at each other's disposal and sharing risks and benefits. With a partnership the aim is to enhance the synergies between the organization and the stakeholders. Partnership tools are, for example, joint ventures, development of projects, multi-stakeholder initiatives and business alliances.

The optimum process of realization of the stakeholder engagement (AccountAbility & United Nations Environment Programme Stakeholder Research Approach, 2005b) is based on a strong commitment by the leaders and on three guide 'principles':

- Importance: what is important for the organization and the stakeholders;
- Completeness in the understanding of all the impacts regarding the activities of the organization and the respective stakeholders' points of view;
- Answering or capability of answering, to guarantee an appropriate answer to the stakeholders' requests and requirements.

For an effective realization of the stakeholder engagement the phases that must be introduced by the university and placed in a strategic perspective are:

- Analysis and planning, with the definition of the strategic objectives of the engagement, the mapping of stakeholders and the clarification of the priorities to follow. In this phase, even in the light of the experiences of other organizations, the margins for improvement are defined, the characteristics of the stakeholders are understood along with their needs and their expectations, while monitoring the progress made. The construction of a matrix can be useful here to synthesize strategic objectives, scopes of intervention and evaluation of the importance, both for the organization and the stakeholders;
- Analysis of organizational capabilities, to strengthen the problem-solving and decision-making capabilities, improving the expertise of the participants in the process, allowing the stakeholders to fully understand the process and to be actively involved in it;
- Drafting of the scheme of the real path of involvement and its realization, which requires the analysis and choice of the tools considered most suited in order to attain information and opinions

DOI: 10.1057/9781137351937

from stakeholders, to improve the quality of the relationship, the trust and the transparency of the communication, to confirm the significance of the chosen theme and to identify possible future actions;
- Reassessment of the actions embarked on and reporting of the results achieved – the final phase, with the goal of planning and reassuring the stakeholders that their contributions will be taken into consideration during the decision-making process.

After the actions in the organizational scope (assigning precise responsibilities in the university and establishing figures, bodies and organizational units dedicated to the relevant themes to ensure control and total co-ordination, supported by the strong imprimatur of the governing bodies) and the broadening of the governance system (with the mapping of stakeholders and the activation of a stakeholder engagement process), the long-term sustainability framework is completed by an appropriate definition of the strategic planning system, which is made concrete in processes and tools through which the university defines objectives and assumes commitments in a long-term perspective, which will then find coherent description in the short term. The strategic approach to sustainability can take on different aspects, which reflect various orientations and motivations, ascribable to four paradigmatic situations:

- Passive strategy, when the university suffers sustainability as a constraint, avoiding including it in its reasoning. In this case the university does not recognize sustainability as its raison d'être, towards which its actions should be directed. The sustainability theme is faced with the perspective of the 'downstream purification', that is acting in retrospect, without, however, introducing changes into the activities carried out;
- Adaptive strategy, when the university begins to become aware of its responsibility in terms of sustainability, but there is not yet an investment of resources in this direction. The university adapts to the requests of the pressure groups, or to legislative provisions, allowing these to define the amount of resources and the action modalities to be dedicated to sustainability;
- Reactive strategy, where the university begins to adopt responsibility in terms of sustainability, reacting to external stimuli, for example providing itself with clean technologies and allocating resources to sustainability governance, but it does so essentially in

'competitive' terms, in comparison with other universities, so as not to lose the capability of attraction in the context of reference;

- Proactive strategy, when the university adopts sustainability as a responsibility, spreading sensitivity towards and awareness of the environmental and social themes among all stakeholders, along the whole organizational pyramid and towards the outside, with a decision-making process permeated by sustainability.

In the strategic-organizational approach the coherent orientation is naturally that of a proactive nature, with the university living and breathing sustainability beyond its boundaries and beyond what is established by legislative provisions. Orienting the strategic planning strategy in a sustainability process means coherently setting out the strategic axes and the objective that the university intends to pursue in the long term.

A strong signal from the university, in terms of the official nature and the irreversibility which attests to its acceptance of sustainability, can be made in the acknowledgement of the relevant concepts in the obligatory documents provided for by the legislative provisions, in the Statute, the Strategic Plan. The introduction of contents (articles) dedicated to sustainability in the Statute constitutes in effect a declaration of commitment by the university, a choice that fits into the governance system. Another important contribution is the construction of the Strategic Plan, in which sustainability represents one of the objectives to pursue and informs the other medium- and long-term institutional objectives, with actions duly described and performance indicators (which will be dealt with later on) that enable them to be monitored over the years.

The ethical code, another tool that typically fits in a perspective of social responsibility, in the case of the university institutions has become obligatory, with the law 240 of 30 December 2010: 'Regulations in the matter of organization of the university, of the academic staff and recuiting, as well as authorization by the government to increase the quality and the efficiency of the university system.' Article 2, Paragraph 4 of the law states: 'the universities are requested to adopt ... an ethical code of the university community formed by academic staff and lecturers, from the technical-administrative staff and from the students of the university'; furthermore, 'the ethical code defines the fundamental values of the university community, fosters the recognition and the respect of individual rights, as well as the acceptance of obligations and responsibilities towards the institution of belonging, and dictates the rules of

DOI: 10.1057/9781137351937

conduct in the interest of the community. The regulations are aimed at avoiding forms of discrimination and of abuse, as well as regulating the cases of conflict of interests or of intellectual property rights'. The presence of an ethical code thus no longer constitutes a special element for the university, defining itself simply in terms of what is requested by law. The university's commitment therefore seems much more meaningful and significant in the case of the voluntary introduction of sustainability contents as the Statute and the Strategic Plan, for which no specific indications are provided for in terms of sustainability contents to be included obligatorily.

In addition to the definition in a sustainability perspective of obligatory documents provided for by the law, proactive behaviour can be made concrete even in the activation of processes and tools of a voluntary nature, which are always set in the long-term perspective. Through these processes and tools the university further confirms its commitment towards sustainability and the assumption of responsibilities in the medium to long term, identifying objectives and actions, usually described for the stakeholders. Even in order to give credibility to the process, it is fundamental that the university periodically activates a monitoring system to verify progress in the commitments adopted, possibly reporting the results to stakeholders, as provided for by the engagement process.

2.2.2 Performance measurement and evaluation system

Performance measurement and evaluation systems in universities have been investigated with contributions that, even though proceeding from different perspectives, considered mainly the traditional scope of action of the university: didactics, research and areas of the administrative services and auxiliaries. In some cases, for the non-profit organizations systematizations and applications of strategic interpretation are proposed, using for example the model of the balanced scorecard (Kaplan, 2001, p. 361), which points out the central role of the mission in orienting company strategy.

A modelling of the performance measurement system based on the sustainability perspective entails the consideration of elements that stress the complexity. The pursuit of sustainable behaviour requires the university to go beyond its own boundaries, in space and in time. This means that, in terms of space, the perimeter of reference is increased,

DOI: 10.1057/9781137351937

that is the different categories of stakeholder that must be considered, not only the subjects directly interested by the core issues (didactics and research), but all those which, even indirectly are affected by the effects of the university's actions.

The time horizon is also extended; the university cannot limit itself to the short-term effects of the activities that are institutionally intended for it, because the effects of a particular behaviour can emerge even at a distance of years, as generally occurs with regard to environmental and social impacts.

The performance measurement system must catch and represent not only the activity carried out, the resources used and the immediate results, but also orient itself towards the long-term impacts produced by the university's intervention.

On the basis of the indicator typologies detailed below, a system based on a strategic-organizational approach also requires a vision integrated among the three constituent dimensions of sustainability (social, economic and environmental). The pursuit of the sustainability objectives cannot be left only to a combination of sectional indicators, each regarding one dimension, but one must also try to introduce intersecting indicators, which simultaneously consider more than one variable.

The methodological approach that is proposed for a performance measurement system compatible with sustainability logics, which coherently fits in the strategic organizational approach, provides for the consideration of indicators linked to the following variables (Mio and Borgato, 2012):

- Process of production and supply of the service;
- Stakeholder category;
- Dimension of the sustainability concerned (environmental, social, economic).

The first variable holds a central role in the evaluation of the total value of the set of indicators in the sustainability perspective, in which, as will be seen, a crucial element is represented by the impact indicators. The result proposed in the present work is made concrete in a scheme of indicators where the various elements that define the total framework in which the action of the university develops are represented, permitting an understanding of the role of each element and any mutual interdependencies.

DOI: 10.1057/9781137351937

In particular, the proposed model provides for the following indicator typologies (Bergamin Barbato and Mio, 2008; Mio and Borgato, 2012):

- Scenario indicators, that is indicators targeted at representing the context of reference. These are indicators that allow the university to understand the characteristics of the context in which it acts, appreciating its dynamics. The degree of 'governability' of such indicators (in the sense of possibility of intervention on the dynamics considered) by the body can be different, depending on the case in question. Scenario indicators can be represented by the entepreneurial attitude of the province or region (or other territory of reference) where the university is present, for example in terms of percentage subdivision by activity sector and respective birth rate, death rate and rate of development. The percentage of selective collection of waste (sustainable refuse and recycling collections) in the province where the university works can also be a scenario indicator, which in that case can constitute a reference for the activation of sustainable behaviour inside and outside the university. Scenario indicators should be considered for a definition of the university's policies even with regard to the specificity of the context and can also constitute objectives to pursue, if expressly targeted. In this case they represent at the same time the starting point and the point of arrival, which is to say the impacts of the interventions introduced, targeted at answering the expectations of individual stakeholders;
- Input indicators, which represent the resources that the university makes use of in the realization of the activity. They can be expressed either in monetary or in non-monetary terms, for example: number of teachers (and relevant cost), number of places available, number of lecture halls, number of libraries, number of volumes, number of electronic periodicals, number of online databases;
- Activity indicators, which allow quantification of the volume of activities realized by the university, for example the number of active courses; number of hours of education provided, number of places reserved for competitions, number of participants in admission exams, number of opening hours of the libraries;
- Output indicators, which express the results of the activity carried out by the university (to be evaluated by comparing objectives vs actual performances), for example: number or percentage of subjects that have obtained educational qualifications, rate of

DOI: 10.1057/9781137351937

withdrawal between first and second year of enrolment, number or percentage of students that did not pass exams, number of subjects of the territory involved in university initiatives;

- Outcome indicators, which express the impact on stakeholders (also) produced by the university actions. These indicators are characterized by strong interdependencies with subjects and systems external to the institution and by reference to a time horizon of the medium to long term. Examples are: satisfaction indexes of the various stakeholder categories (collected through various tools) and the percentage of graduates that have found jobs within a certain period of graduation. The performance of the university in terms of outcome is therefore co-defined by external factors, including economic and cultural aspects, as also demonstrated in some empirical analyses (Jabnoun, 2009).

Besides these, the set of indicators can contain 'need' indicators, which are designed to catch and to represent the needs and expectations of stakeholders. For example, with reference to students or staff with disabilities, the university can identify the specific requirements of these subjects, as in the case of the demolition of architectural barriers, or of services provided for them. The indicator scheme thus constructed therefore intersects two other variables:

- Stakeholder categories, so to succeed in constructing and pinpointing specific sets of issues and metrics, tailored with reference to the stakeholders identified by the university in the mapping process;
- The constituent perspectives of sustainability: economic, environmental and social, to express university performance. Among the numerous contributions proposed for the identification of the environmental, social and economic indicators, the one recognized at an international level as the main reference is the Global Reporting Initiative model (GRI, 2011).

The methodology proposed for the modelling of the system obviously does not exhaust all the indicator typologies. For example, indicators that show the degree of efficiency and of effectiveness of university action can be duly introduced. The scheme can be integrated with indicators that compare resources to results (efficiency), and the results achieved to the objectives (effectiveness).

DOI: 10.1057/9781137351937

With regard to sustainability indicators, a fundamental reference at an international level is constituted by the GRI, a body that defined guidelines to orient organizations in the production of sustainability reports (a theme that will be dealt with in the next sub-section).

A combination of the categories indicated in the methodological approach presented and the typologies of the indicators provided for by the GRI (version G4) is proposed below (Tables 2.1 to 2.6). Supposing that the application of the GRI model to universities in some cases requires a specific reassessment or description, the whole set of indicators provided for are interrelated anyway, since the objective of this analysis is that of highlighting the specific weight of the categories indicated. Furthermore, it must be pointed out that in some cases, the description of the GRI indicator is composite, in the sense that it can accept various cases, representing through a single coding, for example, the activity, the output or the outcome. In this case, in the column 'Category' all the traceable cases in the GRI indicators will be indicated.

TABLE 2.1 *Classification GRI economic performance indicators*

Aspect and related GRI indicators	Category
Economic performance	
EC1 Direct economic value generated and distributed	Input
EC2 Financial implications and other risks and opportunities for the organization's activities due to climate change	Input – Scene
EC3 Coverage of the organization's defined benefit plan obligations	Input
EC4 Significant financial assistance received from government	Input
Market presence	
EC5 Ratios of standard entry-level wage by gender compared to local minimum wage at significant locations of operation	Input – Scene
	Input
EC6 Proportion of senior management hired from the local community at locations of significant operation	Input
Indirect economic impacts	
EC7 Development and impact of infrastructure investments and services supported	Input – Outcome
EC8 Understanding and describing significant indirect economic impacts, including the extent of impacts	Outcome
Procurement Practices	
EC9 Proportion of spending on local suppliers at significant locations of operation	Input - Outcome

DOI: 10.1057/9781137351937

As regards the Economic Performance Indicators, a substantial predominance of input indicators in comparison to the other typologies emerges. Such evidence can seem, on the one hand, physiological in comparison to the nature of the analysed perspective (the economic dimension), in itself probably oriented to represent the resources used. In a sustainability perspective, however, the reference to the economic dimension must go beyond the representation meant in the meaning 'balance of a business year', moving towards the consideration of the general economic impacts in the long term and outside the boundaries of the organization. On the other hand, the representation modalities of the outcomes can be different, because of the difficulties of measurement of the issues, as confirmed later on, being able to be made concrete in a simple description of the phenomenon, up to the quantitative expression of the repercussions generated in the context of reference.

TABLE 2.2 *GRI environmental performance indicators*

Aspect and related GRI indicators	Category
Materials	
EN1 Materials used by weight or volume	Input
EN2 Percentage of materials used that are recycled input materials	Input
Energy	
EN3 Energy consumption within the organization	Input
EN4 Energy consumption outside of the organization	Input
EN5 Energy intensity	Input/Output
EN6 Reduction of energy consumption	Output
EN7 Reductions in energy requirements of products and services	Output
Water	
EN8 Total water withdrawal by source	Input
EN9 Water sources significantly affected by withdrawal of water	Outcome
EN10 Percentage and total volume of water recycled and reused	Input
Biodiversity	
EN11 Location and size of land owned, leased managed in, or adjacent to, protected areas and areas of high biodiversity value outside protected areas	Input
EN12 Description of significant impacts of activities, products and services on biodiversity in protected areas and areas of high biodiversity value outside protected areas	Outcome

Continued

DOI: 10.1057/9781137351937

TABLE 2.2 *Continued*

Aspect and related GRI indicators	Category
EN13 Habitats protected or restored	Activity
EN14 Number of IUCN Red List species and national conservation list species with habitats in areas affected by operations, by level of extinction risk	Scenario
Emissions	
EN15 Direct greenhouse gas (GHG) emissions (Scope 1)	Output
EN16 Energy indirect greenhouse gas (GHG) emissions (Scope 2)	Output
EN17 Other indirect greenhouse gas (GHG) emissions (Scope 3)	Output
EN18 Greenhouse gas (GHG) emissions intensity	Input/Output
EN19 Reduction of greenhouse gas (GHG) emissions	Output
EN20 Emissions of ozone-depleting substances (ODS)	Output
EN21 NO_x, SO_x, and other significant air emissions	Output
Effluents and Waste	
EN22 Total water discharge by quality and destination	Output
EN23 Total weight of waste by type and disposal method	Output
EN24 Total number and volume of significant spills	Output
EN26 Identity, size, protected status and biodiversity value of water bodies and related habitats significantly affected by the reporting organization's discharges of water and runoff	Scene
Products and services	
EN27 , Extent of impact mitigation of environmental impacts of products and services	Activity – Outcome
EN28 Percentage of products sold and their packaging materials that are reclaimed by category	Output
Compliance	
EN29 Monetary value of significant fines and total number of non-monetary sanctions for noncompliance with environmental laws and regulations	Output
Transport	
EN30 Significant environmental impacts of transporting products and other goods and materials for the organization's operations, and transporting members of the workforce	Outcome
Overall	
EN31 Total environmental protection expenditures and investments by type	Input
Supplier Environmental Assessment	
EN32 Percentage of new suppliers that were screened using environmental criteria	Output

Continued

DOI: 10.1057/9781137351937

TABLE 2.2 *Continued*

Aspect and related GRI indicators	Category
EN33 Significant actual and potential negative environmental impacts in the supply chain and actions taken	Input, Activity, Output, Outcome
Environmental Grievance Mechanisms	
EN34 Number of grievances about environmental impacts filed, addressed, and resolved through formal grievance mechanisms	Output

For the environmental dimension, the prevailing categories in the GRI definition are represented by input and output, that is the 'environmental balance' of the organization, intended as a representation of the environmental inputs, the resources 'withdrawn' from the ecosystem and used by the organization for the carrying out the activities and the environmental outputs, the substances, variously expressed, added to the ecosystem in the carrying-out of the activity. Once again, the specific weight of the outcome indicators is relatively reduced, tending more frequently for indicators that represent the activity to be requested.

The social dimension in the GRI model presents a larger expression; therefore, various tables below are related, each corresponding to a thematic issue identified by the aforesaid model.

In the thematic issue 'Labour Practices and Decent Work', outcome indicators are almost absent, while the representation of the input indicators such as resources at the (organic) organization's disposal, or the activity and the output are prevailing.

As in the previous case, even for the thematic issue 'Human Rights' the organization's performance is represented substantially by indicators that are inscribed of the input, activity and output categories, while the outcome perspective is almost absent.

Essentially confirmed, with only two exceptions, is the definition gathered up to now in the GRI model for the social dimension. This is characterized by a performance limited practically in the first phase of the 'process', that is, the consideration of inputs, activities, and output, and therefore being limited to the 'immediate' results of the activity carried out.

DOI: 10.1057/9781137351937

TABLE 2.3 *GRI social performance indicators: labour practices and decent work performance indicators*

Aspect and related GRI indicators	Category
Employment	
LA1	
LA1 Total number and rates of new employee hires and employee turnover by age group, gender and region	Input – Output
LA2 Benefits provided to full-time employees that are not provided to temporary or part-time employees, by significant locations of operation	Input
LA3 Return to work and retention rates after parental leave, by gender	Output
Labor/Management relations	
	Input
LA4 Minimum notice periods regarding operational changes, including whether these are specified in collective agreements	Activity
Occupational health and safety	
LA5 Percentage of total workforce represented in formal joint management–worker health and safety committees that help monitor and advise on occupational health and safety programs	Input
LA6 Type of injury and rates of injury, occupational diseases, lost days, absenteeism and total number of work-related fatalities, by region and by gender	Output
LA7 Workers with high incidence or high risk of diseases related to their occupation	Input – Output
	Activity
LA8 Health and safety topics covered in formal agreements with trade unions	Activity
Training and education	
LA9 Average hours of training per year per employee by gender, and by employee category	Activity
LA10 Programs for skills management and lifelong learning that support the continued employability of employees and assist them in managing career endings	Activity
LA11 Percentage of employees receiving regular performance and career development reviews, by gender and by employee category	Output
Diversity and equal opportunity	
LA12 Composition of governance bodies and breakdown of employees per employee category according to gender, age group, minority group membership and other indicators of diversity	Input
Equal remuneration for women and men	
LA13 Ratio of basic salary and remuneration of women to men by employee category, by significant locations of operation	Input

Continued

DOI: 10.1057/9781137351937

TABLE 2.3 *Continued*

Aspect and related GRI indicators	Category
Supplier Assessment for Labor Practices	
LA14 Percentage of new suppliers that were screened using labor practices criteria	Activity
LA15 Significant actual and potential negative impacts for labor practices in the supply chain and actions taken	Activity – Outcome
Labor Practices Grievance Mechanisms	
LA16 Number of grievances about labor practices filed, addressed, and resolved through formal grievance mechanisms	Actitity - Output

TABLE 2.4 *GRI social performance indicators: human rights performance indicators*

Aspect and related GRI indicators	Category
Investment	
HR1 Total number and percentage of significant investment agreements and contracts that include clauses incorporating human rights concerns, or that underwent human rights screening	Activity
HR2 Total hours of employee training on human rights policies or procedures concerning aspects of human rights that are relevant to operations, including the percentage of employees trained.	Activity – Output
Non-discrimination	
HR3 Total number of incidents of discrimination and corrective actions taken	Output – Activity
Freedom of Association and Collective Bargaining	
HR4 Operations and suppliers identified in which the right to exercise freedom of association and collective bargaining may be violated or at significant risk, and measures taken to support these rights	Input – Activity
Child labour	
HR5 Operations and suppliers identified as having significant risk for incidents of child labor, and measures taken to contribute to the effective abolition of child labor	Input – Activity
Forced or Compulsory Labor	
HR6 Operations and suppliers identified as having significant risk for incidents of forced or compulsory labor, and measures to contribute to the elimination of all forms of forced or compulsory labor	Input – Activity

Continued

DOI: 10.1057/9781137351937

TABLE 2.4 *Continued*

Aspect and related GRI indicators	Category
Security practices	
HR7 Percentage of security personnel trained in the organization's human rights policies or procedures that are relevant to operations	Output
Indigenous rights	
HR8 Total number of incidents of violations involving rights of indigenous people and actions taken	Output – Activity
Assessment	
HR9 Total number and percentage of operations that have been subject to human rights reviews and/or impact assessments	Activity - Output
Supplier Human Rights Assessment	
HR10 Percentage of new suppliers that were screened using human rights criteria	Output
HR11 Significant actual and potential negative human rights impacts in the supply chain and actions taken	Activity - Outcome
Human Rights Grievance Mechanisms	
HR12 Number of grievances about human rights impacts filed, addressed and resolved through formal grievance mechanisms	Output

TABLE 2.5 *GRI social performance indicators: society performance indicators*

Aspect and related GRI indicators	Category
Local communities	
SO1 Percentage of operations with implemented local community engagement, impact assessments and development programs	Output
SO2 Operations with significant actual and potential negative impacts on local communities	Input
Anti-Corruption	
SO3 Total number and percentage of operations assessed for risks related to corruption and significant risks identified	Activity – Output
SO4 Communication and training on anti-corruption policies and procedures	Activity – Output
SO5 Confirmed incidents of corruption and action taken	Activity – Output
Public policy	
SO6 Total value of political contributions by country and recipient/beneficiary	Input
Anti-competitive behavior	
SO7 Total number of legal actions for anti-competitive behavior anti-trust and monopoly practices and their outcomes	Output – Outcome

Continued

DOI: 10.1057/9781137351937

TABLE 2.5 *Continued*

Aspect and related GRI indicators	Category
Compliance	
SO8 Monetary value of significant fines and total number of non-monetary sanctions for non-compliance with laws and regulations	Output
Supplier Assessment for Impacts on Society	
SO9 Percentage of new suppliers that were screened using criteria for impacts on society	Output
SO10 Significant actual and potential negative impacts on society in the supply chain and actions taken	Activity - Outcome
Grievance Mechanisms for Impacts on Society	
SO11 Number of grievances about impacts on society filed, addressed, and resolved through formal grievance mechanisms	Activity – Output

TABLE 2.6 *GRI social performance indicators: product responsibility performance indicators*

Aspect and related GRI indicators	Category
Customer health and safety	
PR1 Percentage of significant product and service categories for which health and safety impacts are assessed for improvement	Output
PR2 Total number of incidents of non-compliance with regulations and voluntary codes concerning health and safety impacts of products and services during their life cycle, by type of outcome	Output – Outcome
Product and service labelling	
PR3 Type of product and service information required by the organization's procedures for product and service information and labelling, and percentage of significant products and services categories subject to such information requirements	Activity – Output
PR4 Total number of incidents of non-compliance with regulations and voluntary codes concerning product and service information and labelling, by type of outcomes	Output – Outcome
PR5 Results of surveys measuring customer satisfaction	Outcome
Marketing communications	
PR6 Sale of banned or disputed products	Output
PR7 Total number of incidents of non-compliance with regulations and voluntary codes concerning marketing communications, including advertising, promotion and sponsorship by type of outcoms.	Output – Outcome

Continued

DOI: 10.1057/9781137351937

TABLE 2.6 *Continued*

Aspect and related GRI indicators	Category
Customer privacy	
PR8 Total number of substantiated complaints regarding breaches of customer privacy and losses of customer data	Output
Compliance	
PR9 Monetary value of significant fines for non-compliance with laws and regulations concerning the provision and use of products and services	Output

For the issue 'Product Responsibility', the application to the university requires a specific description to catch the peculiarities. The outcome category seems more present in terms of its request by the GRI, even if its modality of expression can be different, essentially qualitative, or even with quantitative (non financial) metrics.

In general terms, the analysis demonstrates, with the specifications due to the blind applicability of the scheme to the universities, that the orientation of the GRI model towards the outcome, in terms of performance indicators, is rather limited. The performance represented proves to have a barycentre anchored to a short-term perspective, which aims at gathering and highlighting the resources used, the activities carried out and the immediate results of those activities, while the long-term horizon and the representation of repercussions beyond the organization's boundaries are still not very present.

A key factor that makes a performance measurement system oriented towards sustainability is the acceptance of outcome indicators: these should represent the main goal to which the other indicators coherently refer.

The real sustainability perspective which therefore rests precisely on the orientation towards the outcome and means repercussions in the long term, beyond the organizations boundaries, seems to be at present only an omen, certainly far from being realized, even for the reason of the changes that these would require. In the first place, on the tool and procedural level, stressing the outcome in the set of performance indicators would require a significant re-examination of the definition and the degree of extension of the informative support, with important repercussions in terms of a greater volume of data to handle, organize

DOI: 10.1057/9781137351937

and circulate inside and outside the organization. On the other hand, certainly not less important is the change required on the 'cultural' and organizational front: the perspective of the outcome requires a different mentality and it is extremely challenging overall in terms of the scope of responsibility, which must be broadened so as to include responsibilities that go out from the organization's span of control. For a university to accept outcome indicators in the performance measurement system also means to ovversee issuesnot totally governable and also appreciable only with a vision that relays on the short term. If one thinks of the occupational status of graduates, mentioned previously as outcome, the result does not depend only on how much the university realized in terms of educational and of support services, but also – probably overall – on the personal characteristics of the students and on their expectations, on the social and economic situation of the territory and of the actors such as enterprises that represent the potential work demand, and so on (Mio and Borgato, 2012).

In the same way, to set itself sustainability objectives the university must commit itself to form responsible citizens, which adopt sustainable life-styles to spread the sustainability principles. In the long term, the greater support of the community, which subjects that attended the university are part of and which the university has involved in sensitization initiatives and projects, for sustainability logics in terms of behaviour and relevant inpacts on the environment and on the society is certainly not a phenomenon totally governable by the university.

In the initial phase of the university's path towards sustainability it is plausible that the performance measurement system be composed mainly of input, activity and output indicators, which in a more advanced state can function as a kind of monitoring tool of progress as regards objective impacts and eliminating possible obstacles. For example, a reduced participation of the students in education initiatives and sustainability projects intended for them could be an issue noticeable in the short term, that could hinder the pursuit of the objective of forming responsible citizens and of speading sustainable life-styles in the community.

The scope of the university action is heterogeneous; there are areas where it seems relatively easy to measure the impact generated (impact that is spread over several years)and areas where the degree of complexity is greater. Certainly, the difficulty in constructing outcome indicators increases with the expanding of the horizon and of the perspective of

DOI: 10.1057/9781137351937

reference. With regard to the stakeholder category 'Students', the different moments of measuring the impact can be (Mio and Borgato, 2012):

- During the education period, measuring the degree of satisfaction;
- At the end of the study course, considering the employment placement opportunities, for example through the percentage of graduates that are employed within a certain period of obtaining their degree (for example, one year, three years).
- In the long term, when the impact is assessed, for example, at the 'quality' level of the social capital of the community, with obviously higher difficulties of measuring.

Recognition of the need for a change in the orientation of the university also takes place at the legislative level, where important signals towards a progressive expansion of the perspective are introduced, with traits that are coherent with the sustainability principles and a significant role given to the stakeholder engagement process. With these steps the turnaround is officialized, and considered unavoidable, moving towards a new paradigm, with the possibility of persuading the less sensitive universities to make a greater commitment on this front. Nevertheless, the fact remains that the lever of national university regulations can generate a first push towards the strategic-organizational approach, the proactive nature of which identifies it and makes the full expression impossibile through legislative limitations, due to the difficulties of identifying a universally valid approach because of the enormous differences that characterize universities.

A method of self-evaluation of performance that can be applied in the university context is represented by the Common Assessment Framework (CAF), a 'total quality management' tool developed by the European Institute of Public Administration (EIPA) for the evaluation of the public sector (European Institute of Public Administration, 2013). The pilot version was published in 2000 during the first European Conference on Quality in Lisbon (the first draft dates back to 1998); the latest review was in 2013. The CAF is a model adopted voluntarily, aimed at supporting organizations in the public sector in the use of the quality management techniques and targeted at the continuous improvement of the organizations' performance in its widest meaning, from the satisfaction of stakeholders and efficiency to the achievement of institutionally defined objectives.

In the CAF the direct commitment of the leaders to the 'optimum equlibriums' of the organization is required: it is therefore a managerial

DOI: 10.1057/9781137351937

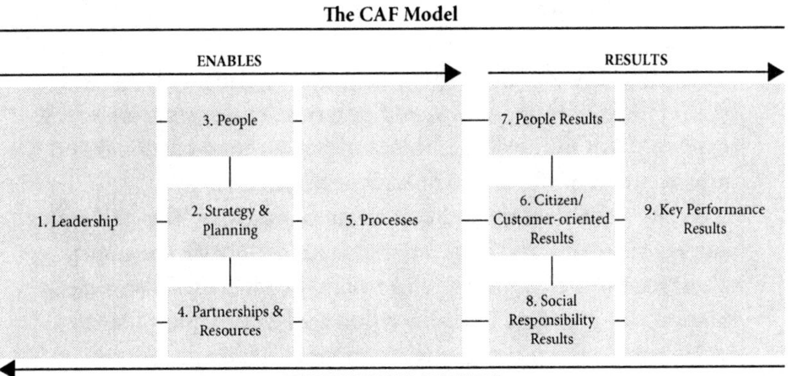

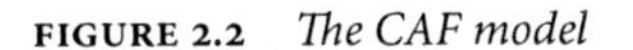
FIGURE 2.2 *The CAF model*

Source: European Institute of Public Administration, 2013, p. 9.

tool that leads towards excellence, identifying the strengths and defects of the organization, in the perspective of continuous improvement.

This self-evaluation model aims at making the relationships of cause and effect in the organization explicit. Its structure comprises five organizational actions or qualifying factors (the enablers) and four scopes of results (the effects).

The evaluation of the results not only concerns the measurement of the organization's performance, but also monitors its utility, that is its capability to resolve or to ease the problems of stakeholders, and the sustainability of the actions embarked on, which means the organization's capability to maintain the interventions put into effect over time.

In the CAF self-evaluation model the importance of stakeholders' feedback is highlighted, as part of the stakeholder engagement process, since these interlocutors can provide valuable information both on the overall perceived image of the university and on the particular relationship between it and the group of stakeholders.

The fundamental pillars of the CAF are eight principles of excellence (European Institute of Public Administration, 2013, p. 11):

1 Principle 1: Results orientation. The organisation focuses on results. Results are achieved which please all of the organization's stakeholders (authorities, citizens/customers, partners and people

DOI: 10.1057/9781137351937

working in the organization) with respect to the targets that have been set;

2 Principle 2: Citizen/Customer focus. The organization focuses on the needs of both present and potential citizens/customers. It involves them in the development of products and services and the improvement of the organization's performance;
3 Principle 3: Leadership and constancy of purpose. This principle couples visionary and inspirational leadership with constancy of purpose in a changing environment. Leaders establish a clear mission statement, as well as a vision and values; they also create and maintain the internal environment in which people can become fully involved in realizing the organization's objectives;
4 Principle 4: Management by processes and facts. This principle guides the organization from the perspective that a desired result is achieved more efficiently when related resources and activities are managed as a process and effective decisions are based on the analysis of data and information;
5 Principle 5: People development and involvement. People at all levels are the essence of an organization and their full involvement enables their abilities to be used for the organization's benefit. The contribution of employees should be maximized through their development and involvement and the creation of a working environment of shared values and a culture of trust, openness, empowerment and recognition;
6 Principle 6: Continuous learning, innovation and improvement. Excellence is challenging the status quo and effecting change by continuous learning to create innovation and improvement opportunities. Continuous improvement should therefore be a permanent objective of the organization.
7 Principle 7: Partnership development. Public sector organizations need others to achieve their targets and should therefore develop and maintain value-adding partnerships. An organization and its suppliers are interdependent, and a mutually beneficial relationship enhances the ability of both to create value.
8 Principle 8: Social responsibility. Public sector organizations have to assume their social responsibility, respect ecological sustainability and try to meet the major expectations and requirements of the local and global community.

DOI: 10.1057/9781137351937

With reference to the measurement of the results ('Results'), in the first three areas (citizens/customers, people, social responsibility) the CAF identifies two categories of indicator:

- Perception measurements, that is indicators aimed at identifying the perception/satisfaction of stakeholders (customers/citizens, staff and society) regarding the activity carried out by the organization. Among the possible areas of the object of such measurements are: the overall image of the organization and its public reputation; the involvement and participation of the citizen/customer in the operational and decision-making process of the organization; accessibility; transparency; the organization's differentiation of services related to the different needs of customers; the layout of the workplace and environmental working conditions; training and empowerment; public awareness of the impact of the organization's performance on the quality of citizens'/customers' lives; perception of the economic and social impact; perception of the approach to environmental issues;
- Performance measurements, that is indicators of the organization's performance. For example: the number of suggestions received and implemented; the opening hours of the different services; the number and processing time of complaints; levels of absenteeism or sickness; rates of staff turnover; response rates to staff surveys; the number of innovation proposals; the number of ethical dilemmas reported, result of social responsibility measurement.

As 'key performance results' the CAF considers two categories of indicator (European Institute of Public Administration, 2013, pp. 51–2):

- External results, which are the measures of the effectiveness of the organization's strategy in terms of its capacity to satisfy the expectations of external stakeholders, in line with the organization's mission and vision. Examples are: results in terms of output (quantity and quality in the delivery of services and products); results in terms of outcome (the effects of the delivered output of services and products in society, and on the direct beneficiaries); results of inspections; and audits on outputs and outcomes;
- Internal results, which are related to the efficiency and effectiveness of internal processes and the economy measures of the working of the organization.

DOI: 10.1057/9781137351937

In general, the self-evaluation proposed by the CAF model confirms the sensitivity to sustainability, projecting itself on the evaluation front of the non-economic performance of the institution, finding basis in the multi-stakeholder approach and identifying sustainability as one of the fundamental objectives.

Another specific contribution was presented in 2003, in collaboration with the association University Leaders for a Sustainable Future. It is a supplementary proposal to the GRI model, which identifies 13 specific indicators, which are optional, to evaluate the educational performance of a university (Lozano, 2006). On the basis of this supplement, a measurement tool of sustainable performance was proposed, the Graphical Assessment of Sustainability in Universities (GASU), which integrates the GRI model (2002 version) with the characteristic dimension of the university, namely its educational and formative dimension, allowing the aspects that the organization excels at and those that are susceptible of being improved to be identified (Lozano, 2011). The set of indicators contains 126 indicators in all, of which 13 are economic indicators, 35 evnvironmental indicators, 48 social indicators and 30 indicators centred on the educational issue, divided in turn into core and additional indicators (Lozano, 2011). Through the use of the indicators characteristic of the university, the educational offer, the research activities and the services offered to students from the sustainability point of view are analysed, evaluating, for example, the number of the research projects and publications on such themes, and the degree to which the concept of sustainable development was integrated within the courses of study.

2.2.3 The reporting system

The reporting of the themes concerning sustainable development, both outside and inside of the university, constitutes an extremely important phase in controlling the sustainability process. Reporting is not meant to be the concluding moment of the process of knowledge and the management of levers, but it also represents, but also represents the modalities through which the university intends to give substance to its sustainability policies. The level of detail selected, the circulation given to the report and the way the report itself is built constitute elements that complete the picture of the decisions to be made on the sustainability theme, which should be co-ordinated with other decisions undertaken in the operational management.

DOI: 10.1057/9781137351937

At the terminological level, there is a distinction between sustainability information and sustainability reporting.

The first, sustainability information, is constituted from the information flow (and as a rule the one produced to support the decision-making process) within the university, which is intended for those that manage the resources (managers). This can take on the shape of a report dedicated solely to sustainability, if it contains specific sustainability information, or it can take shape as a managerial report also including the sustainability themes, among other things.

Sustainability information, therefore, comprises environmental, social and economic information addressed to the managers to support their decision-making processes.

Sustainability reporting, on the other hand, is the totality of the information directed towards the outside. Also in this case, it can take shape as reporting specifically dedicted to sustainability (sustainability report) or be a broader reporting, within which sustainability themes are also dealt with.

In the case of the university, considering the characteristics of this institution in the sustainability perspective, starting from the mission, a planning, programming and therefore also a reporting system on sustainability theme is costitutive and potentially very powerful, overall for the stakeholders. Producing a document such as a sustainability report certainly constitutes an important motivational input towards the acceptance of this paradigm, in the awareness, as already pointed out, that reporting towards the outside must be synergically integrated by coherent interventions on the governance plan, with a stakeholder engagement process, as well as on the performance measurement and evaluation system. In the absence of an overall approach such as the 'strategic-organizational' one, there is a risk that the writing of the report will be perceived by the university community as another fulfilment to satisfy; if this is the case, the report cannot influence the behaviour that characterizes the university's daily acitvity and orient its behaviour. It is a matter of, more significantly, changing the accountability perspective: sustainability seen as a new perspective and an element of innovation in the governing of the university.

Below, sustainability reporting will be focused on outlining the elements of innovation of the external reporting, that is the integrated reporting.

According to the Global Reporting Initiative, 'Sustainability reporting is the practice of measuring, disclosing, and being accountable to internal

DOI: 10.1057/9781137351937

and external stakeholders for organizational performance towards the goal of sustainable development. ... A sustainability report should provide a balanced and reasonable representation of the sustainability performance of a reporting organization – including both positive and negative contributions' (GRI, 2011, p. 3).

The sustainability report is a tool that (Mio, 2002, p. 213):

- Is voluntary, since there is no obligation to write it. It is therefore a totally discretionary reporting initiative activated by the organization;
- Is a free structure, since each organization has absolute discretionary power even in the choice of the form and of the information to present. Currently, different bodies have prepared and published guidelines for writing sustainability reports (the most important at an international level is the GRI), following specific (even on voluntary) frameworks. Generally, the absence of imposed instructions means that sustainability reports is heterogeneous. There are documents that are extremely concise, the fruit of isolated interventions, activated solely for the writing of the report and not founded in formalized procedures, just as, at the extreme opposite, documents with very detailed contents, result in an approach to sustainability reporting that consider external reporting one of the most important tools in external communication;
- Has qualitative and quantitative contents, since, again by virtue of the lack of obligatory instructions on the matter, reports of almost only a descriptive nature can be found, just as reports that provide also information of a quantative nature, of a physical-technical and economic-financial nature. In some reports environmental and social impacts are also represented in monetary terms, through the clarification of costs, investments and environmental and social liabilities. Naturally, the presence of quantitative information and particularly financial information increases the value of the report, attesting to the greater degree of maturity of the company in the field of sustainability. This is particularly valid when the coherent connection with the financial report and also the conceptualization of the balance sheet items are conceived in a sustainability perspective;
- Is periodical: its writing must not be intended as a one-off initiative. Even if it is not an obligatory document, it must not be written occasionally;

DOI: 10.1057/9781137351937

- Is public: it is a reporting tool directed towards the outside, and therefore it must be made accessible so that it can be consulted by stakeholders;
- Is certifiable: that is, susceptible to an affirmation process, intended as the result of a systematic and documented approach, by means of which an independent and qualified professional (or team of professionals) gathers and evaluates the truthfulness of the statements (verifiable) contained in it.

The characteristics of voluntariness and the absence of obligatory instructions with regard to the organization of the sustainability report have brought about the proliferation of documents with extremely different characteristics over time – whence the need for standardization to facilitate the organizing of data collection and preparation of the report, and the comparability among different organizations over time. There have been numerous initiatives undertaken by different bodies with regard to the establishment of standards, with a distinction between process standards and contents standards.

Process standards are focused on the process of organizing a report, that is the co-ordinated, transversal and multidisciplinary totality of activities bent on promoting a change in the organization towards sustainability. This process is periodically monitored and reported through the sustainability report. Among the reference principles in the process standards are AccountAbility's AA1000 series, 'principles-based standards to help organisations become more accountable, responsible and sustainable. They address issues affecting governance, business models and organisational strategy, as well as providing operational guidance on sustainability assurance and stakeholder engagement. The AA1000 standards are designed for the integrated thinking required by the low carbon and green economy, and support integrated reporting and assurance. The standards are developed through a multi-stakeholder consultation process which ensures they are written for those they impact, not just those who may gain from them. They are used by a broad spectrum of organizations – multinational businesses, small and medium enterprises, governments and civil societies' (AccountAbility, 2013).

The AA1000 Series of Standards is organized into three main types (AccountAbility, 2013):

- The AA1000 AccountAbility Principles Standard (AA1000APS), which provides a framework for an organization to identify, prioritize and respond to its sustainability challenges;

DOI: 10.1057/9781137351937

- The AA1000 Assurance Standard (AA1000AS), which provides a methodology for insurance practitioners to evaluate the nature and extent to which an organization adheres to the AccountAbility Principles Standard;
- The AA1000 Stakeholder Engagement Standard (AA1000SES), which provides a framework to help organizations to ensure that stakeholder engagement processes are purpose-driven and robust and deliver results.

Contents standards are directed at defining and standardizing the contents of the report, proceeding from the condition that the report is the result of a process that has a central element in the stakeholder engagement process. The main reference at the international level is represented by the GRI Guidelines: 'The GRI Reporting Framework is intended to serve as a generally accepted framework for reporting on an organization's economic, environmental, and social performance. It is designed for use by organizations of any size, sector, or location' (GRI, 2011, p. 3). The GRI model will be the object of closer examination later on.

There are also other standards, generally focused on specific elements of sustainability, which organizations refer to in drafting their reports, with impacts also at the level of information flows and therefore of organizational processes to protect. They include:

- London Benchmarking Group Model (London Benchmarking Group, 2013), designed to assess the real value and impact of community investment on both business and society. 'The LBG model enables CCI – Corporate Community Investment – professionals to measure their company's overall contribution to the community, taking account of cash, time and in-kind donations, as well as management costs. The model also records the outputs and longer-term community and business impacts of CCI projects';
- SA8000, devised by Social Accountability International, whose mission is to advance the human rights of workers around the world. 'The SA8000 standard is the central document of our work at SAI. It is one of the world's first auditable social certification standards for decent workplaces, across all industrial sectors. It is based on conventions of the International Labour Organisation (ILO) [and] UN and national laws. The SA8000 standard spans industry and corporate codes to create a common language for

DOI: 10.1057/9781137351937

> measuring social compliance. Those seeking to comply with SA8000 have adopted policies and procedures that protect the basic human rights of workers. The management system supports sustainable implementation of the 'elements' of SA8000: child labor, forced and compulsory labor, health and safety, freedom of association and right to collective bargaining, discrimination, disciplinary practices, working hours, remuneration' (Social Accountability International, 2013).

The motivations that have made the GRI Guidelines establish themselves as the main reference at the international level in the realm of sustainability reporting, are mainly two: the first is that it is not limited to only providing a list of contents to include in the report; and the second is that it identifies much more significantly the principles that must inspire the writing: 'The Sustainability Reporting Guidelines (the Guidelines) consist of Principles for defining report content and ensuring the quality of reported information. It also includes Standard Disclosures made up of Performance Indicators and other disclosure items, as well as guidance on specific technical topics in reporting' (GRI, 2011, p. 3).

The strong points of the GRI Guidelines are therefore ascribable to three aspects:

- Completeness, because the methodologies and the indicators provided for refer to all aspects of performance and can be used by any organization of any size, sector or country;
- Structuring, attributed to the minimum requirements provided for, that is information that must be present in the report (the core indicators, mentioned previously), regardless of the size, of the sector or of the geographical position of the company. The 'obligatory' presence of these indicators is to be contextualized, though, insofar as the organization has chosen to adopt the GRI Guidelines, whose application remains voluntary; thus the organization can decide not to include all the core indicators if it decides to write a report only partially in line with the standard (Daub, 2007);
- Flexibility, in that it is necessary to give the right emphasis to the specific characteristics of each sector, guaranteed through the introduction of the additional indicators (a subject dealt with previously).

DOI: 10.1057/9781137351937

The GRI Guidelines provide for two typologies of principles to orient the organization in the drafting of the document: Reporting Principles for Defining Content and Reporting Principles for Defining Quality.

There are four Reporting Principles for Defining Content (GRI, 2011, pp. 7–13):

- Materiality: the information in a report should cover topics and Indicators that: reflect the organization's significant economic, environmental, and social impacts or that would substantively influence the assessments and decisions of stakeholders;
- Stakeholder inclusiveness: the reporting organization should identify its stakeholders and explain in the report how it has responded to their reasonable expectations and interests;
- Sustainability context: the report should present the organization's performance in the wider context of sustainability;
- Completeness: coverage of the material topics and Indicators and definition of the report boundary should be sufficient to reflect significant economic, environmental, and social impacts and enable stakeholders to assess the reporting organization's performance in the reporting period.

There are six Reporting Principles for Defining Quality, the 'guide choices on ensuring the quality of reported information, including its proper presentation' (GRI, 2011, p. 13):

- Balance: the report should reflect positive and negative aspects of the organization's performance to enable a reasoned assessment of overall performance;
- Comparability: issues and information should be selected, compiled, and reported consistently. Reported information should be presented in a manner that enables stakeholders to analyse changes in the organization's performance over time, and could support analysis relative to other organizations;
- Accuracy: the reported information should be sufficiently accurate and detailed for stakeholders to assess the reporting organization's performance;
- Timeliness: reporting occurs on a regular schedule and information is available in time for stakeholders to make informed decisions;
- Clarity: information should be made available in a manner that is understandable and accessible to stakeholders using the report;

DOI: 10.1057/9781137351937

- Reliability: information and processes used in the preparation of a report should be gathered, recorded, compiled, analysed, and disclosed in a way that could be subject to examination and that establishes the quality and materiality of the information. (GRI, 2011, pp. 13–17)

The contents of a report written according to the GRI Guidelines are indicated in the section Standard Disclosures, which specifies the base content that should appear in a sustainability report (GRI, 2011, p. 19).

There are three types of disclosure indicated in the GRI framework (GRI, 2011, p. 19):

- Strategy and Profile: disclosures that set the overall context for understanding organizational performance such as its strategy, profile, and governance;
- Management Approach: disclosures that cover how an organization addresses a given set of topics in order to provide context for understanding performance in a specific area;
- Performance Indicators: Indicators that elicit comparable information on the economic, environmental, and social performance of the organization [which were presented previously].

In spite of the spread of the sustainability report, there are numerous critiques expressed with reference to the issues as the standards used in the writing of the report, the degree of coherence of the information, the verifiability of the data disclosed, and even 'the actual' strategy of the organization.

Specific contributions are really critical even to the fundamental of the sustainability report, but really the criticism is referred to the applications and to the companies procedures. In fact, despite methodological supports are now available for the evaluation and reporting of environmental and social variables, organizations tend not to make use of them, thereby not providing an effective and balanced development and integration of the three constituent dimensions: economic, social and environmental (Gray and Milne, 2002; Gray, 2010). The application of the Triple Bottom Line perspective is considered unsatisfactory, so the report is realized as a prospect where the prime enhancement is in the economic aspect (the variable that is predominant), to which limited information of a social and environmental nature is added, making

DOI: 10.1057/9781137351937

impossible a complete evaluation of the existing relations among the three perspectives (and relevant conflicts) and an appreciation of the final result. This means recognizing that the sustainability is a global concept (beyond the Triple Bottom Line), which emphasizes not only the efficent allocation of resources, but also the equity of their distribution among the various subjects, as well as among the current generation and the future ones. To report its degree of sustainability the organization must make a complete and transparent declaration on the way in which it contributed, in a positive or negative way, to world sustainability. This entails the production of a complex and detailed analysis of the interactions of the organization with the ecosystem, its resources and the community, which must be supported by a considerable quantity of data.

Another criticism concerns the role that the sustainability report often has for the organization: not that of reporting with transparency, completeness and truthfulness to stakeholders the company's policy and its performance from the economic, social and environmental point of view, but that of reporting information duly and instrumentally selected in order to improve its reputation (Bebbington et al., 2008). Naturally the characteristics of the sustainability report outlined previously affect this: from its voluntary nature to the freedom in the form and in the contents, which leave organizations discretionary power over the information they transmit to the outside, possibly also omitting the negative aspects, and a declaration of adherence to the GRI Guidelines obviously cannot be translated in a real certification of the quality of the report. Of extreme criticality is, consequently, the theme of assurance of the sustainability report, which impacts on the control and monitoring of the completeness and on the reliability of the data used. In fact, only for specific issues an assurance could be provided (typically quantitative and verifiable data); in case of the descriptive and qualitative information a specific metric should be provided, trying to considering all different stakeholders. As a consequence only a limited assureance would be provided with negative repercussions on the information and the importance of the report for stakeholders (Ballou et al., 2006).

With reference to the scope of the universities, the sustainability report is certainly a fundamental tool to periodically report the effects of responsible governance and its correspondence to declared strategies. Furthermore, if written through a participatory process, the report can also constitute the prerequisite for sharing with stakeholders, the university's values and policies, as well as being a tool that allows sufficient

DOI: 10.1057/9781137351937

understanding and satisfaction of the stakeholders' requirements through the engagement process. Given the ever greater importance attributed to the involvement of stakeholders and to the intrinsic complexity of the mapping of stakeholders inside the university, it is fundamental that the sustainability report clearly reports the processes and the actions taken in order to educate the stakeholders as to their active involvement (Hinna, 2002).

The transparency in the sharing of the information, the involvement and the consequent satisfaction lead to a more extensive and incisive social legitimization of the university and of its conduct in its landscape. In the university, legitimization by stakeholders has a great value by virtue of the remarkable number of subjects the university relates with, besides the role that it is institutionally called to carry out. In this sense, reporting concentrated substantially on economic-financial information would limit the sharing of information to 'economic' stakeholders, excluding others with interests (Hinna, 2002).

Even for universities the sustainability report is a voluntary document, whose writing is not regulated by legislation, depending mostly on the sensitivity and motivation of the governing bodies, of their perception of the governance system in a sustainability perspective and therefore on their broadening to the different categories of stakeholders. At present, even though compared to the project sector, it's still relatively less intense and frequent both in terms of the evolution of university structures and behaviour towards sustainable development (Albrecht et al., 2007; Fonseca et al., 2011), important signals are picked up at the level of the increase in the number of universities that write a report aimed towards the outside (Locatelli and Schena, 2011).

The writing of the sustainability report enables universities to increase their degree of transparency, by making information not disclosed before available in the perspective of engaging stakeholders in an effective dialogue. The collecting process of the necessary data for the writing of the report, which often proves to be extremely demanding, especially the first time, enables the university to broaden its information base, since new information is collected and analysed. In many cases the most demanding challenge is precisely that of managing to collect information 'resident' in different parts of the organization and systematize it, adapting it to the new context, transforming the knowledge of individuals into shared knowledge that can be encoded and transferred. In this

DOI: 10.1057/9781137351937

sense, the sustainability report can be a tool for memorizing knowledge (Albrecht et al., 2007). The writing process of the report should follow a procedure, establishing and sharing unambiguously the typology of data and information to collect with the owners at each different step of the process (all aspects obviously should be adjusted from year to year).

For the universities, the stakeholder engagement process can find its moment of activation precisely with the writing of the sustainability report, which can serve as the opportunity to start sharing a path, positively exploiting the links with the community, as well as in the international research community (Becher and Kogan, 1992). The different stages in the writing process, from the collection of the data (some of which are available for the first time) and the analysis and sharing of the same up to the definition of the objectives can constitute driving factors to trigger a necessary evolution of the university, at the level of sustainability culture, at the organizational level, with potential positive effects on the management system and on the motivation of the employees.

Some concerns can be raised with regards to the cost of a high quantity of data and the possible misinterpretation of the information, overall for the environmental and social aspects, whose understanding appears complex even as related toa combination of concomitant causes and a long term horizon (Albrecht et al., 2007).

With reference to guidelines for the writing of the report, the GRI framework is among the most used by universities (Lozano, 2011), being the milestone of several industries. However it provides a wide range of applicative indications, it points out the significant failings on the part of the organization and permits an easier comparison with other sectors (Fonseca et al., 2011). Difficulties of application of the GRI model to the university remain, though, in spite of the integration of numerous social indicators previously dedicated exclusively to the civil service. Such difficulties are ascribable to the typical characteristics of the organizational structure of universities, the high interdependency among their activities and above all to the 'educational' dimension that characterizes them (Lozano, 2006). For these reasons, in 2011, fewer than 30 of the world's universities wrote a sustainability report in accordance with the GRI Guidelines, a fact that confirms the low incidence of sustainability reporting in the university sector (GRI, 2012).

From an analysis carried out (Lozano, 2011) on 12 universities all over the world it emerged that in the report universities usually give importance first to environmental aspects and second to economic ones,

DOI: 10.1057/9781137351937

essentially neglecting social aspects. Some reasons could be, for the environmental dimension, the greater obviousness in some countries of environmental deterioration compared to social degradation, while for the economic dimension, the more immediate availability of economic data, which is required for the preparation of obligatory economic-financial reports. For the social variable, often the data presented concern work activities, without stressing adequately the commitment lavished by the university on themes such as community well-being and human rights.

Other empirical facts (Fonseca et al., 2011) confirm the need to develop processes and tools not only of evaluation but also of sustainability reporting in universities. The evaluation tools must be comprehensible to the majority of stakeholders and the reporting must be clear and verifiable; if the information appears not easily accessibile and understandable on the part of stakeholders, the result and the value of the tools are perceived as limited. In particular, from the research it emerged that the students and the bodies responsible for sustainability inside universities perceive as ever more pressing the need to understand and report, inside and outside, the sustainability performance of the university and that furthermore the explicit introduction of the sustainability theme in the courses of study and in the research projects (core issues of the university) attains greater importance at a media level, and therefore greater visibility, than the actions concentrated on other support services offered by the university. Elements of criticality were found, besides the lack of a proper assurance of the report, also in terms of the scarce visibility given to reports on university websites, which could be interpretated as not coherent with the declaration of commitment by the leaders towards sustainability presented in the report or included in the universities' mission.

Even other researchers confirm the difficulty of easily accessing the report which is published in detail on-line (for the reduction of printing costs and the consequent consumption of paper), although less frequently in a paper version. In addition the timeliness is criticized too. (Locatelli and Schena, 2011).

In sum, from the several studies carried out on the sustainability report in universities (Fonseca et al., 2011; Locatelli and Schena, 2011) emerges a common need for improvement in reporting, at several levels, including the greater diffusion of the tool and of the process sustainability reporting among universities, the suitable balance among the three aspects of

DOI: 10.1057/9781137351937

sustainability, the level of disclosure of the information, and the respect of generally accepted standards or frameworks even for educational performance.

Besides the necessary improvements in the application of the sustainability principles and in the writing of the sustainability report, there is always an urgency for the universities to put into practice the 'walk as you talk' principle, so as to be an example and stimulus to all stakeholders and to students in particular, since university action is concentrated on them to shape them into responsible citizens.

Although at present the sustainability report is a relatively little used tool among universities, in the last few years the external reporting front has seen the start of the conceptualization of a new framework, integrated reporting (IR), which 'brings together material information about an organisation's business model, strategy, governance, performance and prospects in a way that reflects the commercial, social and environmental context within which it operates. It aims to communicate the "integrated thinking" through which management applies a collective understanding of the full complexity of value creation to investors and other stakeholders' (International Integrated Reporting Council, 2013). While the sustainability report usually joins the traditional financial report (obligatory) and therefore has no choice but to be a tool more concentrated on environmental and social aspects, leaving less space for the economic dimension already dealt with in the obligatory report, the integrated report should become a concrete reality as the solution aligned to the sustainability paradigm, since it aims at pursuing a real integration and balance among all the dimensions, besides projecting the same report in the long term, as a result of an effective orientation in that timescale of the organization. 'At the heart of IR is the growing realization that a wide range of factors determine the value of an organization – some of these are financial or tangible in nature and are easy to account for in financial statements (e.g. property, cash), while many are not (e.g. people, natural resources, intellectual capital, market and regulatory context, competition, energy security). IR reflects the broad and longer-term consequences of the decisions organizations make, based on a wide range of factors, in order to create and sustain value. IR enables an organization to communicate in a clear, articulate way how it is drawing on all the resources and relationships it utilizes to create and preserve value in the short, medium and long term, helping investors to manage risks and allocate resources most efficiently' (IIRC, 2013).

DOI: 10.1057/9781137351937

The journey towards the definition of the framework of integrated reporting is extremely complex and challenging, since it is a matter of providing guidance and principles not so close to a the traditional perspective of the organizations management (IIRC, 2013):

- Strategic focus: an Integrated Report provides insight into the organization's strategic objectives, and how those objectives relate to its ability to create and sustain value over time and the resources and relationships on which the organization depends;
- Connectivity of information: an Integrated Report shows the connection between the different components of the organization's business model, external factors that affect the organization, and the various resources and relationships on which the organization and its performance depend;
- Future orientation: an Integrated Report includes management's expectations about the future, as well as other information to help report users understand and assess the organization's prospects and the uncertainties it faces;
- Responsiveness of stakeholder inclusiveness: an Integrated Report provides insight into the organization's relationships with its key stakeholders and how and to what extent the organization understands, takes into account and responds to their needs;
- Conciseness, reliability and materiality: an Integrated Report provides concise, reliable information that it's material to assessing the organization's ability to create and sustain value in the short, medium and long term.

The same 'content elements' provided for by IR identify a report with definitely different characteristics compared to the definition known and adopted by the organizations up to now:

- Organizational overview and business model;
- Operating context, including risks and opportunities;
- Strategic objectives and strategies to achieve these objectives;
- Governance and remuneration;
- Performance;
- Future outlook.

In the case of the university, the proposed framework would certainly require a specific description to allow for the specificities compared with other types of organizations. Yet, the core elements of integrated

DOI: 10.1057/9781137351937

reporting could constitute the target towards which the universities' actual reports progressively shift, in a strategic organizational approach towards sustainability.

2.3 Carbon management: guidelines for universities

The contents of this chapter are extracted from the document *Guidelines on the Matter of Carbon Management for Universities* (Ministry of the Environment and of the Protection of the Territory and of the Sea, University Ca' Foscari Venice, 2012), one of the outputs of a pilot project, and part of a Memorandum of Understanding between the Ministry of the Environment and of the Protection of the Territory and of the Sea (MATTM) and Ca' Foscari University of Venice (the Carbon Management Project is illustrated in Chapter 3).

2.3.1 General concepts

In all European countries, public and state institutions have embarked on new policies and new strategies to reach carbon reduction objectives and to conform to the emissions reduction obligations dictated by international and European Community regulations. The two great thematic axes are: the control of energy requirements and of other resources; and the fulfilment of energy requirements through the use of sources that replace fossil fuels. Since the main environmental objective is the reduction of greenhouse gas emissions, the indicator chosen for the definition of the objectives and for the monitoring of the results over time is carbon, from carbon dioxide, or CO_2, one of the gases mainly responsible for the greenhouse effect. Thus the definition of carbon managemenent, is the following: carbon management refers to the policies, procedures and systems dedicated to the management of such variable as a whole.

The reasons for adopting a carbon management policy in any state or private institution extend well beyond environmental issues. As it is easy to sense, the positive effects in terms of better management of energy consumption are, besides environmental, also of an economic nature, thanks to the money saved on energy bills, although these are subject to continuous cost increases. In the case of state-owned instutions or enterprises, other reasons are added to the one already described that are related to the organization's positioning and leadership; organizations

DOI: 10.1057/9781137351937

must give the numerous stakeholders collateral evidence of their performance, supplying with quality indicators for the service provided.

To achieve good results, it is crucial to demonstrate a clear 'vision' in terms of the management of the carbon footprint and aligned with international standards and best practice: it is fundamental that in universities the carbon management policies do not end up as an addition to the existing activities, but that they are rather the fruit of a gradual transformation due to an integrated approach within the existing organization.

An efficient carbon management policy introduced in the institutions of higher education contributes to the improvement of economic wealth and, as a consequence, to the reduction of total state expenditure, thanks to the reduction of waste and to the identification of investment opportunities. The costs saved this way can be set aside for educational and research projects. Carbon management brings educational and economic, in addition to environmental, advantages: if one takes into consideration market forces and the probable future increase of energy costs (caused by the reduction of the availability of hydrocarbons and by the regulations for the reduction of emissions), carbon management also becomes synonymous with good risk management and better planning. In this way, future increases in energy costs and new regulations will have a negative impact that will be presumably mitigated.

2.3.2 Carbon management process

The best activities of carbon management are inspired by the Deming Cycle, based on the sequence Plan-Do-Check-Act (PDCA).

Translated into practical terms, it is possible to break down carbon management into a succession of main steps:

1 Adoption of the carbon management strategy: the first step is obtaining the commitment of the decision-making bodies (management, rectorship, board of governors, university senate, senior management), which must demonstrate that they share the principles and the aims of the adoption of a carbon management policy. This step is connected to motivational reasons and they are made explicit through subsequent actions (the adoption of a Carbon Policy, the establishment of a process management commitee, and the establishment of a Carbon and Energy Management team, in charge of following the process from the tecnical-operational point of view);

DOI: 10.1057/9781137351937

2 Definition of the baseline: once the commitment of the decision-makers is obtained and the technical team is set up, it is necessary to define the starting point, the so-called baseline. A current statement of emissions and consumption. The identified baseline is the fixed point of reference for the definition of the improvement objectives; in the case of carbon management, the baseline is supplied by the carbon footprint of the organization. It is calculated according to the international reference standard (ISO 14064 part 1) and, to give it more validity, it should be certified by an independent body;
3 Identification of the improvement measures: the team must implement suitable actions to identify and quantify the improvement measures;
4 Once the measures are identified it is necessary to draw up a plan for their implementation, in order to obtain the approval of the Management Committee;
5 The final step is constituted by following the operational phases for the realization of the measures identified and of all the activities for reporting the results, all of which are carried out in the perspective of the iterative cycle that restarts and then integrates with step 2) described above.

The main aspects related to points 2–5 of the process listed above will now be presented in succession, with a particularly in-depth analysis of the calculation of the carbon footprint, while the treatment of the strategical aspects, which is the first point of the process, will be dealt with in Chapter 3, described with specific reference to Ca' Foscari's experience.

2.3.3 The calculation of the carbon footprint

One of the definitions of carbon footprint is 'quantity of CO_2 and other greenhouse effect gases discharged by a particular activity or organization' (Carbon Trust). More precisely, the carbon footprint represents the amount of greenhouse gases generated through the whole life cycle of a good/service or the activities of an organization during a certain period.

The carbon footprint of a university represents the management target, that is the baseline for the creation of sustainable carbon management. One of the first and most important steps that concerns the carbon management of a university is the calculation of the carbon footprint. The carbon footprint includes the greenhouse gas emissions related to the whole

DOI: 10.1057/9781137351937

life cycle of the university services (including energy consumption, goods purchased, staff travels, the mobility of students and employees, and waste management). These emissions are classified in the Scope 1, 2, 3 (described below) and are initially calculated for a defined solar year (baseline *year*). The choice of the baseline is made by looking for the most recent year for which data are sufficiently complete and coherent. The data for the reference year must be reliable and of high quality in order to supply a valid and significant comparison and to guarantee the transparency and repeatability of the calculations, and the object of a subsequent certification.

One of the methodological pillars of reference for the calculation of the carbon footprint is the Life Cycle Assessment (LCA), which considers the environmental impact of the goods/services during the whole life cycle, in order to adopt the appropriate corrective and mitigating actions in all the phases of the lives of the products, from their production to their final disposal. The LCA methodology, up to today, has been regulated by ISO regulations of the series 14044 approved at the international level and in force since 1997. Subsequently, approaches were established that focus on the theme of climate change, with the definition of other standards and methodologies related to the carbon footprint, that is ISO 14064. The first part of the ISO 14064 regulation draws on a previous work, the Corporate Accounting and Reporting Standard – GHG Protocol, an international standard to calculate and communicate the emissions developed by the World Resources Institute (WRI) together with the World Business Council for Sustainable Development (WBCSD). The GHG Protocol, launched in 2001 and revised and reissued in 2003, is, even today, the computation tool and the (carbon) reporting standard that is the most widespread at a worldwide level. The GHG Protocol was created following a global and inter-departmental in-depth analysis that involved firms, NGOs and governments and for that reason it is considered the output of a multilateral consensus iregarding calculation and verification of the greenhouse gas emissions. In the course of 2009, the WRI and the WBCSD also developed two new GHG Protocol standards: the Scope 3 Accounting and Reporting Standard, and the Product Life Cycle Accounting and Reporting Standard.

Another reference standard is the Publicly Available Specification (PAS) 2050, derived from the ISO 14044 standard and developed jointly by the UK's Department for Environment, Food and Rural Affairs (DEFRA), Carbon Trust and British Standards Institute (BSI). It supplies

DOI: 10.1057/9781137351937

a method standard for the calculation of the carbon footprint of goods/services, focusing exclusively on the greenhouse gas emissions during the life cycle of a product, and considers the emissions associated with the whole chain of a product/service.

The evaluation of a carbon footprint must be informed by the principles of credibility, transperency and coherence present within the methodology WRI/WBCSD GHG Protocol Corporate Standard.

According to those set out by the Kyoto Protocol, for the calculation of a carbon footprint the emissions of the six types of gas mainly responsible for the global warming effect must be evaluated: carbon dioxide (CO_2), methane (CH_4), nitrous oxide (N_2O), hydrofluorocarbons (HFCs), sulfur hexafluoride (SF_6) and perfluorocarbons (PFCs).

The unit of measurement for the calculation of a carbon footprint is the CO_2 equivalent. It summarizes the influences of the different greenhouse gases on global warming, coherently with international standards. The contribution of each gas undergoes a process of normalization through a specific index called Global Warming Potential – GWP.

Before calculating the carbon footprint of a university, it is necessary to clearly define the operational and organizational boundaries within which one will work (system boundary.

The operational boundary of a university includes all the university's activities that produce greenhouse gas emisssions; they are precisely described in the GHG Protocol and can be grouped into three scopes: Scope 1, Scope 2 and Scope 3.

The greenhouse gas emissions of Scope 1 (direct emissions of greenhouse gases) are those generated directly by sources possessed by the university or under the university's control. In the case of a university Scope 1 includes:

- Emissions deriving from the stationary combustion of fossil fuels for heating and other fuels to generate electricity on site;
- Emissions deriving from the combustion of fossil fuels during the use of vehicles owned/controlled by the body;
- 'Fugitive' emissions resulting from the intentional or non-intentional release of greenhouse gases, including refrigerant leaks (HFCs) from cooling equipment and the release of CH_4 from the breeding of animals, laboratories, etc;
- Emissions deriving from other agricultural activities, should they be present;

DOI: 10.1057/9781137351937

- Other emissions directly ascribable to the internal implications ne of the university activities.

The greenhouse gas emissions of Scope 2 (indirect greenhouse gas emissions under the direct control of the institution) are mainly those generated by the production of electricity used by the institution and also by the heating/cooling purchased (both from the national grid and from the local grid).

The greenhouse gas emissions in Scope 3 (indirect emissions of greenhouse gases not directly controlled by the institution) include all indirect emissions not comprised in Scope 2 that are a 'consequence of the institution's activities, but come from sources that are not possessed or are not controlled by the institution'. These include the mobility of students and employees; employees' travel; the disposal of waste; emissions from the extraction, production and transportation of raw materials purchased; international students' trips and those related to summer schools; leaks during the transmission and distribution of electricity; fugitive emissions during the extraction of fuels; the transportation and distribution of fuels used for the activities of Scopes 1 and 2; and other outsourced activities. In general, the emissions in Scope 3 are difficult to measure, but since they contribute an important amount to the inventory of greenhouse gas emissions of the university, it is advisable to include them in the inventory from the start.

The emissions in Scope 3 can be subdivided into three categories: upstream emissions, downstream emissions and other emissions.

Upstream emissions are those that occur in the life span of the inputs until the point of receipt by the university (cradle to grave methodology). The inputs include: supply of goods and services, transportation and distribution of supplied goods; emissions related to the energy not covered in Scope 2; investment capital; waste created in the upstream operations of the institution. Downstream emissions are those generated in the life span of the outputs of the university. They are linked to the disposal of waste generated by the university, staff missions, student and staff commuting, air travel of international students and transportation between the university and the residence of international students.

The organizational boundary of the university is defined according to two main assumptions:

- All university activities within its geographical borders are considered;

DOI: 10.1057/9781137351937

- All the buildings owned and/or under direct control of the university are taken into account as well as the buildings in which the university is responsible for the payment of the electricity bills.

After the scopes of the greenhouse gas emissions that are to be included in the system boundaries have been identified, the next step is the collection of all the necessary data to carry out the calculation of the carbon footprint.

To achieve a good result it is advisable to draw up an accompanying methodology note in which all the measures and steps related to the calculation procedure are clearly defined. The calculation method is divided into three stages corresponding to the three emission scopes defined by the GHG Protocol (according to the standard 'the organizations will report and represent separately the scopes 1 and 2') with the purpose of being able to represent, also in graphic form, the effect of the three different scopes.

2.3.4 Identification of the objectives of emission reduction

The identification of the objectives for the emission reduction is a strategic element for the university; the targets identified will have to consider the economic and social management policies implemented or to be implemented and pay close attention to the technical aspects of feasibility. A list of measures designed to reach the planned reduction is proposed below:

- To respect the legislative obligations dictated by national and international bodies;
- Starting from the first analysis of the real estates, to improve energy efficiency with the aim to achieving a reduction of emissions into the atmosphere;
- To replace non-renewable sources with renewable sources for the production of energy needed for the management of the buildings;
- In cases where it is necessary to choose among non-renewable sources, to promote, in addition to the economic aspect, also the technical aspect in order to guarantee the lower use of primary energy;
- To manage the technological equipment with the objective of reducing to a minimum the energy wasted and to guarantee conditions of hygrothermic well-being for the occupants;

DOI: 10.1057/9781137351937

- To carry out systematic and programmed maintenance on the building equipment systems in order to maximize the energy performance of the properties;
- To aim to obtain, even in part and in the long term, the self-sustainability of the properties from the energy point of view.

2.3.5 Definition of a carbon management plan and monitoring

The carbon management plan (CMP) of the university is a document that outlines in detail the university stategy for the reduction of carbon dioxide emissions from its products (goods or services) in a specified time span. The CMP is the main result of the carbon management team (CMT) activities and represents the necessary background to the implementation of a reduction strategy for emissions and for the energy use of the university. This implies producing specific documents subject to the approval of the management commitee and that periodically undergo updating depending on the development of the plan.

The CMP is specific for each organization and reflects its situation depending on the regional requirements, both climatic and socio-cultural. The main objective of the CMP is to lay out, in the most precise and detailed manner, the way to reach a reduction in emissions. To reach this objective it is therefore advisable that all the phases above (carbon policy, identification of the objectives, definition of boundaries and baseline, computation of the carbon footprint) are concluded before the drawing up of a good CMP.

2.3.6 Reporting plan

Proper and timely reporting of the project actions, addressed to both students and staff, is a key factor in ensuring the maximum effectiveness of this important process of change. It is thus advisable to prepare a reporting plan accurately structured and integrated in the general CMP. A reporting plan should generally include a synthesis of the planned actions, their realization time and the responsibilities identified. This means that the key elements of the reporting plan are the selection and individuation of: active and passive parties (stakeholders), themes to confront, messages to spread, suitable means of reporting, action timing, allocation of responsibilities, involvement/engagement activities and diffusion of good practices.

It is important that the reporting activity is constant over time and that it reports the actual results obtained with the reduction actions deployed

DOI: 10.1057/9781137351937

in the CMP so as not to cause a spot effect that has little effectiveness and that would be perceived as a mere sign of intentions that are not followed by actual actions. Considering that a correct implementation of the CMP is also based on evident support of the institutional leaders and on widespread commitment from all human resources of the university, it is desirable that the information on the execution of the CMP, the progress of the staff awareness campaign and the specific activities of carbon management are communicated to all the people responsible for the buildings.

2.4 Green universities

In the path towards sustainability of universities, the consideration and acceptance of the environmental dimension probably presents a greater degree of innovation than other constituent dimensions, due to theinherent mission of universities, projected more directly on the sociality and the need for economic constraints as determined by legislative obligations.

In this view, the commitment to the environmental variable by universities has led to the conceptualization 'green university', intended to connote the universities' commitment on the sustainability front, with particular reference to the environmental dimension (for example, implementing projects for carbon management and green campuses).

The theme of the green university is essentially connected to two main scopes: the networks established for universities that are commited on the sustainability front and the UI GreenMetric World University Ranking. These aspects will be the object of treatment in the following paragraphs.

2.4.1 Sustainability networks

Numerous initiatives and networks have been established in order to unite the organizations that have decided to adhere to the sustainability paradigm, with approaches that are more or less broad. Among these, some are specifically dedicated to institutions that operate in the field of education and in particular to universities: Principles for Responsible Management Education (PRME, mentioned previously) and ISCN and Greenmetric.

DOI: 10.1057/9781137351937

The PRME network has, besides the framework composed of the six fundamental principles, which serve as orientation for the institutions that adhere to it, defined the Engagement Model for PRME Schools & Academic Institutions (Principles for Responsible Management Education, 2013): 'The PRME can serve as a framework for systemic change for business schools and management-related institutions, on the basis of three distinctive characteristics of the initiative:

1 Continuous Improvement: any school that is willing to engage in a gradual but systemic manner is welcome to join the initiative. Implementation of the Principles should be understood as a long-term process of continuous performance improvement and the PRME can provide a framework of general principles through which to engage faculty and staff, and build institutional support;
2 A Learning Network: the PRME initiative also functions as a learning network. By collecting and channeling good practices, it will facilitate an exchange of existing and state-of-the-art experiences within the PRME network;
3 Report to Stakeholders: adopting the PRME implies that the signatory school is willing to report regularly – annually – on progress to all stakeholders. Public reporting is the best way to ensure the credibility of the initiative and to allow for public recognition for good performances.'

The model outlines the characteristics of implementation of the framework, expressing clearly the gradual but systematic approach through which institutions must accept the six fundamental principles (continuous improvement), emphasizing the importance of a virtuous loop/circle connected to the effects of learning and the sharing of best practice that must be activated among the institutions of the network. In this network, reporting is a necessary step in the governing process of sustainability, to give credibility to the process using the reporting to the stakeholders about activities and results.

The network is also particularly active in the involvement of universities in initiatives, projects and working groups concentrated on the closer examination of the specific aspects of the sustainability. Participation in working groups enables among the various aspects the institutions involved to acquire greater visibility on the front of the commitment

DOI: 10.1057/9781137351937

towards sustainability. Over time working groups with a focus on different aspects have been set up, including:

- Working Group on Anti-Corruption in Curriculum Change;
- Working Group on Gender Equality;
- Working Group on Poverty, a Challenge for Management Education;
- Working Group on Sharing Information on Progress;
- Working Group on Sustainable Leadership in the Era of Climate Change;
- Working Group on the Incorporation of the Principles in Executive Degree Programs;
- Working Group 50+20 – Management Education for the World Joint Project.

In addition to the working groups, PRME Regional Chapters have been created, whose task is essentially that of describing the issues in context to the network and to sustainability at a local level, considering the specificities of a particular area (Principles for Responsible Management Education, 2013). In particular, the PRME Regional Chapters must:

- Provide a platform for dialogue, learning and action based on responsible management and leadership education and research;
- Increase the visibility of PRME and its signatories in a region;
- Adapt the Six Principles of PRME to the local context;
- Develop and promote activities linked to the Principles.

PRME Regional Meetings are also important for the visibility of the network and the spreading of sustainability principles in the member organizations. In order to maximize the effectiveness and to unify the organizational aspects of the meetings, as well as giving a common imprint that is recognizable at an international level, a specific contribution was organized 'Guidelines and Recommendations for PRME Regional Meetings', which organizations must follow when they evaluate the opportunity to host the event.

The International Sustainable Campus Network (ISCN), founded in 2007, 'provides a global forum to support leading colleges, universities, and corporate campuses in the exchange of information, ideas, and best practices for achieving sustainable campus operations and integrating sustainability in research and teaching' (International

DOI: 10.1057/9781137351937

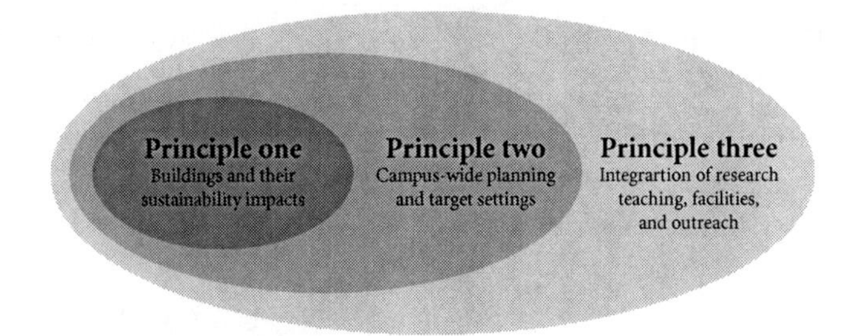

FIGURE 2.3 *Hierarchy of principles – key focal points of international exchange*
Source: International Sustainable Campus Network, 2013

Sustainable Campus Network, 2013). The most prestigious universities adhere to the network, including Stanford, Harvard, Yale, Cambridge and Oxford.

The three principles on which universities should concentrate on the path towards sustainability, as indentified by the ISCN, are shown in Figure 2.3.

The pursuit of these three principles is supported by the network through a series of actions and initiatives that fit into a general strategy (International Sustainable Campus Network, 2013):

- ISCN-GULF Sustainable Campus Charter. Institutions may sign the charter and commit to setting their own, concrete targets against the shared Charter principles mentioned above, and to report transparently on their progress against those targets;
- Working Groups dedicated to each of the three Charter Principles conduct research and facilitate the development of resources to support knowledge exchange;
- Conferences and Symposia are held across the globe to address the whole picture of campus sustainability or focus more closely on particular issues of strategic relevance for campus sustainability;
- Sustainable Campus Excellence Awards are given out annually to highlight best practices and provide public recognition to campuses excelling in campus sustainability.

Of particular importantce is the ISCN/GULF Sustainable Campus Charter, whose signing 'represents an organization's public commitment

DOI: 10.1057/9781137351937

to aligning its operations, research, and teaching with the goal of sustainability. The signatories commit to:

- implementing the three ISCN/GULF sustainable campus principles described later on,
- set concrete and measurable goals for each of the three principles, and strive to achieve them,
- and report regularly and publicly on their organization's performance in this regard' (International Sustainable Campus Network, 2013).

In the document the three principles mentioned previously (Figure 2.3) define clearly a sustainable campus, principles that the signatory organizations commit themselves to pursue:

1. Principle 1: To demonstrate respect for nature and society, sustainability considerations should be an integral part of the planning, construction, renovation and operation of buildings on campus;
2. Principle 2: To ensure long-term sustainable campus development, campus-wide master planning and target-setting should include environmental and social goals;
3. Principle 3: To align the organization's core mission with sustainable development, facilities, research and education should be linked to create a 'living laboratory' for sustainability.

With the three principles a sort of progressive sensitivity towards sustainability is outlined: from the focus on the buildings, the action becomes more widespread, going on to involve planning and target-setting and eventually the mission, with the intention of making the university a 'living laboratory' for sustainability. To support universities in the acceptance of these principles the ISCN drafted the Implementation Guidelines to the ISCN-GULF Sustainable Campus Charter, in which indications are provided, to orient the institutions that decide to adhere to the Charter and to translate the principles into actions, as well as report to the outside the commitment undertaken. (Charter Reports).

As well as networks of global importance, such as those just described, connected initiatives have gradually become established; these are always voluntary and are of a more limited nature. Some of these aim at defining a sustainability ranking among the institutions involved. This is the case, for example, with the United States Green Report Card initiative,

DOI: 10.1057/9781137351937

which 300 universities adhere to; it takes the form of a grading system based on sustainability information collected among the participating universities. The result is expressed in qualitative terms, with the assignation of a judgement that is ranked on a scale (A to F). While valuable, the initiative still presents weak points in its 'local' application and in the relatively small number of universities involved, along with the qualitative modality of expression of the sustainability ranking.

This scope instead has been tackled by a network that has acquired importance at a global level, UI Green Metric, also in regards to the connected definition of the UI GreenMetric World University Ranking, an initiative of Universitas Indonesia launched in 2010. 'As part of its strategy of raising its international standing, the University hosted an International Conference on World University Rankings on 16 April 2009. It invited a number of experts on world university rankings.... It was clear from the discussions that current criteria being used to rank universities were not giving credit to those that were making efforts to reduce their carbon footprint and thus help combat global climate change' (UI Green Metric, 2013). The initiative of the network and the related ranking is the result of the acquired awareness of the importance of the sustainability themes (in particular for the environmental dimension) among universities, along with the activation of sustainability projects by various top world universities, among them Harvard, Chicago and Copenhagen, in addition to projects that concern smaller universities. The overwhelming adoption of the sustainability strategy in the universities clashes with criteria and methodologies adopted for the definition of the university rankings, in which the sustainability is defintively not included.

The following sub-section is dedicated to this concern.

2.4.2 University ranking and sustainability

The growing importance of sustainability themes at a general level and for universities in particular has led to the creation of sustainability rankings, including the already mentioned UI GreenMetric World University Ranking. The university that devised the initiative, Universitas Indonesia, 'saw the need for a uniform system that would be suitable to attract the support of thousands of the world's universities and where the results were based on a numerical score that would allow ranking so that quick comparisons could be made among them on the criteria of their commitment to addressing the problems of sustainability and

DOI: 10.1057/9781137351937

environmental impact' (Universitas Indonesia, UI, Green Metric, 2013). This initiative has two meanings, supporting each other: to contribute to the cultural debate on sustainability in the universities and to change the management inside the universities.

Regarding the first contribution, 'it is expected that by drawing the attention of university leaders and stake holders, more attention will be given to combating global climate change, energy and water conservation, waste recycling, and green transportation. Such activities will require change of behavior and providing more attention to sustainability of the environment, as well as economic and social problem related to the sustainability' (UI Green Metric, 2013).

On the other hand, universities recognize increasingly the commitment to sustainability, under external pressure and stakeholders actions. There is also the risk of affecting the reputation of the institutions if universities do not follow this new strategy and do not move towards new models of production and consumption.

The initiative of Universitas Indonesia provides a big push towards these aims, publishing the results of an online survey: 'regarding the current condition and policies related to Green Campus and Sustainability in the Universities all over the world', which concludes with the assignation of a total score. This contributes to disclosing information to universities that demonstrate attention towards sustainability themes (in particular for the environmental aspect) and have started up projects and actions in this sense, with best practices outlined. In a broader perspective, the intention of Green Metric to activate a virtuous circle, taking care that the results of the ranking constitute inputs for the universities (and their governing bodies) for a more effective acceptance of sustainability logics, 'to put in place eco-friendly policies and manage behavioral change among the academic community at their respective institutions' (UI Green Metric, 2013).

From its first edition in 2010 the methodology for the definition of the ranking has evolved, with a perspective of reference focused on three Es: Environment, Economics and Equity.

The types of information requested have been selected among those considered important, in general terms, for the universities as regards the sustainability, for example, the size of the university and its zoning profile, whether urban, suburban or rural; the amount of green space; its electricity consumption, because of its link to the carbon footprint; transportation; its water usage; and its waste management processes. Furthermore, information targeted at outlining the university's

DOI: 10.1057/9781137351937

commitment to sustainability on the level of policies and actions as well as on the reporting front, are requested. The issues considered in the latest available ranking (2012) and the respective weight used for determining the total score, are the following:

- Setting and Infrastructure (SI), 15%;
- Energy and Climate Change (EC), 21%;
- Waste (WS), 18%;
- Water (WR), 10%;
- Transportation (TR), 18%;
- Education (ED), 18%.

The definition points out a prevailing aim on the enviromental dimension, while the social dimension remains at the bottom.

The indicators considered for the purpose of the 2012 ranking are explained below, classified according to the categories of the methodology illustrated above,.

The first area, 'Setting and Infrastructure' provides 'the basic information of the university consideration towards green environment. This indicator also shows whether the campus deserves to be called a Green Campus. The aim is to trigger the participating university to provide more space for greenery and in safeguarding environment, as well as the development of sustainable energy' (UI Green Metric, 2013).

TABLE 2.7 *Classification of 'setting and infrastructure' indicators*

Setting and infrastructure	Category
1. Campus setting	Input
2. Type of higher education institution	Activity
3. Number of campus sites	Input
4. Total campus area (square metres)	Input
5. Total ground floor area of buildings (square metres)	Input
6. Number of students	Activity
7. Number of academic and administrative staff	Input
8. Area of campus covered in vegetation in the form of forest (provide as percentage of total site area)	Input
9. Area of campus covered in planted vegetation (include lawns, gardens, green roofs, internal planting) (provide as percentage of total site area)	Input
10. Retention: non-retentive surfaces on campus as percentage of total area (where non-retentive surfaces incl. earth, grass and con-block, and retentive surfaces incl. concrete and tarmac) for water absorption	Input
11. Percentage of university budget for sustainability effort	Input

DOI: 10.1057/9781137351937

The area 'Energy and Climate Change' has the highest weight: 'with this indicator, universities are expected to increase the effort in energy efficiency in their buildings and to care more about nature and energy resources' (UI Green Metric, 2013).

Analyzing the area 'Waste', it is clear that 'waste treatment and recycling activities are major factors in creating a sustainable environment. The activities of university staff and students in campus will produce a lot of waste; therefore some programs and waste treatments should be among the concern of the university, that is a recycling program, toxic waste recycling, organic waste treatment, inorganic waste treatment, sewage disposal' (UI Green Metric, 2013).

TABLE 2.8 *Classification of 'energy and climate change' indicators*

Energy and climate change	Category
1. Energy efficient appliances usage	Input
2. Renewable energy resources	Input
3. Electricity usage per year	Input
4. Energy conservation programme	Activity
5. Green building elements	Input
6. Climate change adaptation and mitigation programme	Activity
7. Greenhouse gas emission reduction policy	Activity

TABLE 2.9 *Classification of 'waste' indicators*

Waste	Category
1. Recycling programme for university waste	Activity
2. Toxic waste recycling	Activity/Output
3. Organic waste treatment	Activity/Output
4. Inorganic waste treatment	Activity/Output
5. Sewage disposal	Activity/Output
6. Policy to reduce the use of paper and plastic on campus	Activity

TABLE 2.10 *Classification of 'water' indicators*

Water	Category
1. Water conservation programme	Activity
2. Piped water	Input

DOI: 10.1057/9781137351937

TABLE 2.11 *Classification of 'transport' indicators*

Transport	Category
1. Number of vehicles owned by institution (buses and cars)	Input
2. Number of cars entering the university daily	Activity
3. Number of bicycles found on campus on an average day	Activity
4. Transportation policy designed to limit the number of motor vehicles used on campus	Activity
5. Transportation policy designed to limit or decrease the parking area on campus	Activity
6. Campus buses (availability of buses for journeys within the campus, whether free or paid)	Input
7. Bicycle and pedestrian policy on campus	Input/Activity

TABLE 2.12 *Classification of 'education' indicators*

Education	Category
1. Number of courses related to the environment and sustainability	Activity
2. Total number of courses offered	Activity
3. Total research funds dedicated to environmental and sustainability research	Input
4. Total research funds	Input
5. Number of scholarly publications on the environment and sustainability	Activity
6. Number of scholarly events related to the environment and sustainability	Activity
7. Number of student organizations related to the environment and sustainability	Activity
8. Existence of a university-run sustainability website	Activity

'Water use in campus is another important indicator in Greenmetric. The aim is that universities can decrease water usage, increase conservation program, and protect the habitat' (UI Green Metric, 2013).

Another critical theme for universities is that of mobility: 'Transportation system plays an important role in the carbon emission and pollutant level in university. Transportation policy to limit the number of motor vehicles in campus, the use of campus bus and bicycle will encourage a healthier environment. The pedestrian policy will encourage students and staff to walk around campus, and avoid using private vehicle. The use of environmentally friendly public transportation will decrease carbon footprint around campus' (UI Green Metric, 2013).

The last area considered, 'Education', is aimed at identifying the university's commitment to sustainability n terms of the offer and realization

DOI: 10.1057/9781137351937

of teaching and research activities, along with other initiatives of raising awareness and reporting.

In overall terms, besides the above-mentioned concentration on the environmental dimension, the 2012 methodology of the UI GreenMetric World University Ranking leads to the identification of the ranking using almost only two categories of indicators, that is input and activity. The resulting indicators are present solely in the Waste area, as far as the typology of the outputs (and anyway the results are essentially on the qualitative level), while no outcome indicators were found. Therefore it is a ranking that, at least in its current definition, aims to identify primarily the degree of the university's commitment towards sustainability, in terms of the resources dedicated to it and the activities carried out, with a limited focus on the results achieved.

The definition adopted appears to be different from the sustainability indexes for businesses, such as the Dow Jones Sustainability Index, which are characterized by a greater breadth of the areas looked into, as well as by greater concentration on results (or outcomes), as well as in the definition of measurable targets.

The approaches adopted for the purpose of the definition of the 'general' university ratings and of the criteria provided for at the legislative level (for instance for the allocation of resources) are usually characterized by over-simplification. The decisive aspects, such as the size of the universities, choices made in terms of the mix of disciplines and faculties (which determine the balance between research and didactics), connections with the territory of belonging, in addition to sustainability in general, are ignored. In fact, various bodies publish rankings and classifications according to parameters specified each time; but all these ranking tend to look at universities preferring the comparability rather than the identities and the specific charachteristics of each university. A proper performance measurement and evaluation system should be intended as a process for improving strategy and results, not simply as a tool to allocate resources. At present, the most common university rankings present elements of criticality because they focus on partial measures and are potentially distorting and in this regard they could even undermine the university's long term success, as a fundamental actor in the creation of value for the community and society. Think, for example, of the effectiveness of the education process, identified often only as the 'average years needed to obtain a degree': this value could be optimized reducing the quality of education supplied and this is not in the public interest.

DOI: 10.1057/9781137351937

(Mio and Borgato, 2012). Growing sensitivity on sustainability themes requires a reconsideration of the more general rankings, for an inclusion of sustainabilit by all universities. This inclusion is a consequence of recognizing sustainability as a core issue of the university and not as an 'accessory activity'. Rankings dedicated specifically to sustainability are important for the recognition and enhancement of the commitment of universities, contributing to gaining attention on these themes. But they must necessarily tend towards the coherent acceptance in the scope of general rankings. This is unavoidable if the sustainability perspective is to rise to become a fundamental dimension of the university mission, which all universities must consider, because they are vocationally oriented in such a sense, being included in the university agenda and not being considered as a sectorial interest. The inclusion of sustainbility inside the architecture of the general ones could contributeto the spreading in universities of the strategic organizational approach to sustainability; and the only really rewarding way for univesrities and stakeholders for the future.

DOI: 10.1057/9781137351937

3
Ca' Foscari – Case Study

Abstract: *Chapter 3 illustrates Ca' Foscari's path towards sustainability and its fundamental stages, in particular: the strategic organizational approach adopted; the actions; the tools and the projects activated – in particular, sustainability governance, with the acceptance of sustainability in the Statute and in the Strategic Plan; the appointment of organizational figures; the Ethical Code; the mapping of stakeholders and the stakeholder engagement process; the sustainability tools, with the Sustainability Commitments Charter, the Carbon Management Project, the Sustainability Report and sustainability teaching and research; other projects and initiatives, including the introduction of selective collection of waste and the commitment to integrate sustainability into university rankings.*

Mio, Chiara. *Towards a Sustainable University: The Ca' Foscari Experience.* Basingstoke: Palgrave Macmillan, 2013. DOI: 10.1057/9781137351937.

DOI: 10.1057/9781137351937

3.1 Ca' Foscari and the path towards sustainability

The origins of Ca'Foscari University go back to 1868, when by Royal Decree no. 4530 on 6 August, the Royal Secondary School of Commerce of Venice (Regia Scuola Superiore di Commercio di Venezia) was founded.

The University adopted the title Ca' Foscari University (an abbreviation of 'Casa Foscari', Foscari being the name of the family that owned the building) with the Presidential Decree of 28 May 1968, which replaced that of Academic Institute of Economy and Commerce, obtained by the Royal Decree of 1 October 1936.

Ca' Foscari is the oldest business school in Italy and the second oldest in Europe, a university where more than 30 languages are taught. Like the city that houses it, it is a crossroads of cultures, fields of research and ideas. It is a modern university abreast of the times which, in its eight departments, realizes advanced research projects and innovative study programmes to guarantee an educational offer of excellence – an educational offer that, with its six Interdepartmental Schools, Ca' Foscari has made transversal and integrated among several disciplines, so that whoever studies in Venice can acquire broader and more complex competences: economics incorporates cultural heritage, languages run into economic studies, restoration meets the sciences, to give only a few examples.

The university offers courses at all levels, both in Italian and in English: 15 three-year Bachelor's degrees, 30 Master's degrees and 32 Professional Masters Programmes in addition to the Doctorate courses within the

Studying at Ca' Foscari is not only going to lecture halls and libraries. It is not only receiving an education of excellence. It is living a unique experience, in a context that no one else puts together: art, culture, international sensitivity and exchanges with foreign countries.

It is simply Venice.
Welcome to Ca' Foscari, welcome to Venice.

Carlo Carraro,
Rector of Ca' Foscari

FIGURE 3.1 *Welcome to Ca' Foscari*
Source: Ca' Foscari University, 2013b

DOI: 10.1057/9781137351937

Doctorate School. To these are added the new Summer school's courses for credits and extra-curricular activities for Italian and foreign students and professionals interested in courses and lessons given by Italian and foreign professors.

Ca' Foscari University also offers a rich portfolio of international programmes, with five Bachelor degrees, thirteen Master's programmes, two Professional's Master's programmes and ten PhDs (Ca' Foscari University of Venice, 2013b).

Besides the eight departments and six Interdepartmental Schools, seven university libraries are part of Ca' Foscari, giving students the opportunity to consult 700,000 books and over 400,000 periodicals, as well as the support of an extensive network of personal computers and workstations.

TABLE 3.1 *Ca' Foscari at a glance*

Students	
18,939	Students enrolled at Ca' Foscari
More than 4,000	First-year students per annum
3,551	Students getting a degree in the year 2011
More than 700	New graduates participating in the awarding of three-year degree diplomas in Saint Mark's Square
About 3,000	People that have access to the restricted area in Saint Mark's Square
Staff	
151	Full professors
179	Associate professors
2	Professors with temporary appointments
177	Lecturers/University researchers with permanent appointments
20	Lecturers/University researchers with temporary appointments
1	University assistant
7	Directors/Managers
526	Administrative employees
69	Collaborators and language experts
Services and events	
More than 1 million	Publications availabe in the university libraries
2,700	Periodicals on subscription
More than 8,300	Eletronic periodicals
176	Specialist databases
About 800	Cultural/scientific events every year

Source: Ca' Foscari University, 2013a.

DOI: 10.1057/9781137351937

In structural terms, Ca' Foscari is made up of 36 settlements, in 88 buildings, for a total area (internal and external) of 104,863 square metres, of which 80% represents the walkable area. The size of the university buildings varies between 80 m^2 and 7,429 m^2, distributed as follows:

- Smaller buildings, with areas less than 500 m^2, represent 38% of the total;
- Buildings with areas between 500 m^2 and 3,000 m^2 represent the majority, that is 56% of the buildings in question;
- Only 7% of the buildings have an area greater than 3,000 m^2.

Certainly the university stands out for its unique position in the world, which is the city of Venice.

Over time the university has integrated itself into the socio-economic context of the Italian north-east, thanks to numerous and precious relationships with the hetergeneous entrepreneurial and social reality of the community. The city of Venice has characteristics profoundly different from any other in the world, at the level of its territorial morphology and of its infrastructures and structures, which affect the life and the activities of the people living, working and studying there. Certainly also the uniqueness that distinguishes the city of Venice, where the majority of Ca' Foscari's adminstrative, organizational and teaching structures are located, has acted as a propelling factor for Ca' Foscari's embarking concretely on the challenging path of sustainibility, well before the other Italian universities.

There are various fields that require a sustainable approach: from morphologic aspects of the territory that affect the mobility of the people and goods, with problems that concern also the collection and handling of waste, to the protection of the town's precious artistic heritage, a source of world media exposure and an attraction for the millions of tourists who visit Venice every year. Venice also shows its vulnerability due to the high water phenomenon, making the consideration of such themes as environmental and social sustainability in everyday life evident and pressing all those who live and work in Venice (including institutions) to play a decisive role.

The challenge of the preservation and protection of the town, due to the difficult conditions that characterize the environmental and social context where Ca' Foscari operates, has been taken on by the university, which has adhered to the paradigm of sustainable development, starting an important path of interiorization of the cornerstones of sustainability and its spread over the territory, making itself the protagonist in the community.

DOI: 10.1057/9781137351937

In 2008 Ca' Foscari took the first steps towards sustainability, when it volountarily approved an Ethicalal Code, two years before law 240 of 30 January 2010 became effective, which imposed on universities the obligation to adopt this instrument. In the following year, 2009, the Venetian university, with a new rectorship, began to implement important innovations within the theme of sustainability, aiming for systematic action, shared, among employees at the strategic, managerial and organizational levels, proceeding proactively on the path oriented towards responsible governance (Locatelli and Schena, 2011).

Among the first choices characterizing the path embarked on was the establishment, in 2009, of a new post, the Rector's Representative for Environmental Sustainability and Social Responsibility. A member of the university's academic staff, the Representative, who operates on instructions from the Rector, is responsible for the sustainability of the university, co-ordinates activities and projects related to environmental and social sustainability, constitutes the initial contact point for internal and external stakeholders is encharged with the drafting of the sustainability report (Locatelli and Schena, 2011).

Sustainability issues are not dealt with in a sporadic manner, nor driven by single measures or interventions. The approach adopted is inspired by a strategic-organizational approach, which comes from the underlying strategic policy and influences the governance system and all the operational mechanisms.

The adherence to models of environmental and social responsibility requires strong interventions at the level of the value system of the university, besides the construction and implementation of managerial models, which are impossibile to make in the absence of consistent commitment from the entire structure, above all in the initial phase, in which the paths for an effective rationalization of resources, in the perspective of the satisfaction of all the stakeholders are identified (Locatelli and Schena, 2011). At Ca' Foscari a strong imprimatur by the top management and an involvement of the structures from below takes place, with a converging top-down and bottom-up action. The approach adopted by Ca' Foscari's governing bodies for the adoption at a strategic-organizational level of the sustainability perspective developed at the same time:

top down, with significant endorsement and input by the top management, with consistent choices of an organizational nature mainly bent on guaranteeing the maximum consideration/prestige of the interventions,

DOI: 10.1057/9781137351937

bottom up action involving of the entire structure, beginning with significant interactions from the research centers present within the University, extending to all the organizational units and to the various categories of stakeholders, first of all the students; the perspective of sustainability in fact is a 'traversal *asset*' for the University, pervading all fields and identified as a constituting part of the same.

Since 2010 the fundamental steps have proceeded firmly, without interruption or doubts about the implementation of the sustainability strategy of the Venetian university (Table 3.2).

Firstly the key processes and key tools of the university institution are influenced; in the first place, the decision to intervene on the Statute, the 'constitution' of the University. The complex revision process of this tool is started, with regard to the need to conform to the provisons of law 240 of 30 December 2010, which reforms the university system. In 2011 the new University Statute becomes effective, in which the principles of sustaniability are stated (article 3, paragraph 3) and the Ethicalal Code (article 52) and the Sustainability Commitments Charter (article 53) are

TABLE 3.2 *Sustainability step by step*

8 July 2010	Approval of first **Sustainability Commitments Charter** (2010–12)
23 July 2010	Agreement with the Ministry of the Environment for the pilot **Carbon Management Project**
24 November 2010	Start of the Sustainability Report project **Selective Collection of Waste at Ca' Foscari**
27 May 2011	Endorsement by the Board of Governors of the **Strategic Plan**, in which sustainability is one of the ten strategic objectives
8 July 2011	Publication of the **First Sustainability Report** (2010) and approval of the **Sustainability Commitments Charter** (2011–13)
14 July 2011	New agreement "**Addendum**" with the Ministry of the Environment in the scope of the Carbon Management Project
23 September 2011	Publication of the **Guidelines for Carbon Management in Italian Universities**
2 October 2011	The new **University Statute** with the inclusion of the sustainability principles (art. 3.3) and of the Sustainability Commitments Charter (art. 53) becomes effective
27 January 2012	Approval of the new **Sustainability Commitments Charter** (2012–14)
9 July 2012	Publication of the **Second Sustainability Report** (2011)
21 December 2012	Approval of the **Sustainability Commitments Charter** (2013–15)

Source: Ca' Foscari University, 2013c.

DOI: 10.1057/9781137351937

quoted. Ca' Foscari is the first university in Italy to have approved a statute in line with the reform of the university system, where aspects regarding sustainability find room, as well as consolidation of the process of stakeholder engagement, for example, with the PhD students' consultation group and the participation of the representatives of the administrative technical staff to the University Senate and to the Board of Governors of the University. These aspects made Ca' Foscari recognizable as a 'fair, sustainable, meritocratic' University (Ca' Foscari University, 2011a).

Another fundamental action that testifies to the persistence, systematic nature and profundity of the approach adopted by Ca' Foscari is the approval by the Academic Senate and the subsequent publication of the Sustainability Commitments Charter, which is currently the only such charter in the national university context. It is a fundamental constitutive element of the approach of Ca' Foscari towards sustainability, a tool of elevated strategic and operational importance, where the objectives of the university on the themes of economic, social and environmental sustainability are espressed clearly, emerging as a precise adoption of responsibility in terms of the commitments undertaken and declared to stakeholders. In particular, in the Commitments Charter 'the objectives aiming at minimizing the impact of the university on the environment and on the natural resources, at increasing the social cohesion and at reducing the inequalities, fostering the cultural growth and the sustainable economic growth of the territory' are defined (Ca' Foscari University, 2012b, p. 5). The Charter has a three-year validity, and it is updated yearly and systematically monitored monthly or quarterly. It contains strategic and operational strategies, actions, temporal targets and management aids in the ten areas of intervention identified in the course of the planning process, carried out jointly by the management and by the professors.

The Statute and the Commitments Charter are thus the cornerstone tools for the governance of Ca' Foscari, with which the commitment of the university to sustainability is further strengthened and asserted through the enunciation of fundamental principles and values, the definition of objectives and the allocation of precise responsibilities.

A further action was the strategic planning process and its output, the Strategic Plan. After the formalization of the process of allocating responsibilities on the thematics in the Commitments Charter, sustainability also became part of the university's Strategic Plan, an aspect that emphasizes Ca' Foscari's determination to integrate sustainability into all

DOI: 10.1057/9781137351937

its activities, proceeding from a strategically targeted approach. In the Strategic Plan 'Towards Ca' Foscari 2018', the themes of economic, social and environmental sustainability find a place, through numerous references, with one of the ten objectives specifically concentrated on and dedicated to sustainability.

The attention to sustainability and to the stakeholder is expressed in the university's mission, which applies to all the academic disciplines and activities organized inside the university (Clugston and Calder, 1999), using expressions clearly showing a strong commitment to sustainability (Fonseca et al., 2011). The processes and the tools listed above developed by Ca' Foscari constitute the reference framework in the sustainability perspective, moving from driving factors that increase motivation in the people to a progressive dissemination of such concepts, of the principles that are at the heart of sustainability (Calder and Clugston, 2003). Such processes coherently join other tools and initiatives, which will be object of in-depth analysis in the following sections, such as the Sustainability Report and the main projects started by the university on the themes of economic, social and environmental sustainability.

In July 2011 Ca' Foscari publicly presented the report for the year 2010: the Venetian university was therefore the first in Italy to write a sustainability report, where the economic, social and environmental perspectives are presented in a single document. Even at the international level, there are not many universities that write a sustainability report or a social or environmental balance sheet, in addition to the financial balance sheet of the business year and to the financial report required by law (Locatelli and Schena, 2011). The publication of a sustainability report should represent, for any organization, public or private, including university institutions, the concluding moment of a long and complex procedure. The reason lies in the fact that, despite the availability of guidelines and methodologies at the national and international levels, the organization in very few cases has the knowledge, procedures and tools for collecting and analysing sufficient and relevant data to proceed with the elaboration of information of an environmental and social nature.

In the scope of the process of stakeholder engagement, a constitutive element not to be ignored in a strategic-organizational approach such as that of Ca' Foscari, various projects have been started. Among those aimed at involving students is the Selective Collection of Waste at Ca' Foscari project, started in 2010 thanks to a collaboration with the firm that handles the collection of waste in the Province of Venice (Veritas

DOI: 10.1057/9781137351937

S.p.A.). There are two essential elements to the project: the first relates to the rationalization of waste collection on the university premises, considering its particular location in a city substantially without roads or possibilties to make use of road haulage, i.e. a context where the collection of waste is carried out door to door and by boat, resulting in an extremely complicated and costly process. The second project is addressed to internal stakeholders and the local community and it is represented by the promotion of more sustainable daily behaviours. The project has seen a strong appreciation and involvement by the firm that handles the waste and the active participation of students, with various awareness initiatives and communications centred on the objective 'zero waste'.

But the prime initiative to strengthen sustainability in universities and contribute to affirming Ca' Foscari as the reference university on sustainability themes was the Carbon Management Project, a pilot project also started in 2010 that was part of a Memorandum of Understanding with the Ministry of the Environment and for the Protection of the Territory and of the Sea (MATTM). The joint project is aimed at the planning and implementation of an effective model for the calculation of the carbon footprint, particularly difficult in complex structures such as the Venetian university, and the subsequent proposal of corrective measures aimed at reducing the emissions detected. Another output of the project was the publication of the Guidelines for Carbon Management in Italian Universities, which represents a starting point for other Italian universities that want to adopt a management system for CO_2 emissions, supporting the choice of the methodologies and technologies. The strategic aim of the carbon management project concerns both economic aspects, with the rationalization and reduction of the expenditure, and the pursuit of an eco-friendly positioning of the University. In operational terms, the objectives are represented by the computerized management of energy requirements, by the systematic recording of energy consumption, by the implementation of relief measures for the environmental impacts measured and gathered and by the modernization of infrastructures in the pursuit of a higher level of energy efficiency. The collaboration with the MATTM was then renewed with the definition of a new agreement called 'Addendum', aimed at obtaining the certification ISO 14064–3:2006 for the proposed methodology for the calculation of the carbon footprint and, once again, at the behavioural level, for the identification and implementation of

DOI: 10.1057/9781137351937

innovative actions aiming at promoting the adoption of responsible life styles. The 'CO_2 calculator' is part of these, a tool of strong educational value, conceived and made available to students and staff to allow each one to quantify the carbon emissions generated by their daily activities, with the aim of increasing awareness of the importance of behaviour that respects the environment.

In addition to the projects indicated above, which gained considerable attention in the media, Ca' Foscari has started numerous other actions aimed at promoting and spreading awareness and expertise on sustainability themes. Among these was the opening and the management of a website dedicated to sustainability in Ca' Foscari, which can be accessed via a prominent link on the homepage of the university's institutional site. The site represents an essential tool for spreading information and news on Ca' Foscari's sustainability activities, while also facilitating management of the information flow coming from stakeholders (Hinna, 2002). Furthermore, as of the academic year 2012–13, the university enabled students to insert, in their study plan, irrespective of the department to which they belonged, the CDS (Sustainability Expertise) course, corresponding to 1 CFU (University Educational Credit), with the objective of dealing with sustainability themes through a multidisciplinary approach (Shriberg, 2002).

3.2 The governance of sustainability

The governance of sustainability at Ca' Foscari finds concrete expression in aspects of an organizational nature, with the creation of significative elements, at the regulatory and procedural level, such as the Statute, where the governance rules are described in a sustainability perspective, as well as in aspects of a procedural and instrumental nature, with the purpose of supporting the path towards sustainability, marked by the definition of commitments and objectives towards stakeholders.

The choices of governance of sustainability are inserted coherently in a framework that, from the vision, is described in the value system of the Venetian university.

The value system of Ca' Foscari Venice includes an affirmation of its 'secular nature, pluralist and free of any ideologic, confessional, political or economic conditioning' (Ca' Foscari University, 2012a, p. 3). The

DOI: 10.1057/9781137351937

> To become a University that is able to involve all its protagonists in an academic, cultural and professional experience without equal; that marries quality research and excellent teaching methodology to contribute to the innovation and development of its territory and country, and to be recognized as one of the best in Europe.

FIGURE 3.2 *Ca' Foscari's vision*
Source: Ca' Foscari University, 2012a, p. 8.

university 'requests each member of the Ca' Foscari community in the fulfilment of their duties and with regard to the roles and to the responsibilities assumed, to respect, protect and promote the cornerstone values which Ca' Foscari is guided by: protection of human dignity; rejection of every 'unfair discrimination'; enhancement of the merit and of the individual and cultural diversities; respect of freedom and of basic rights; support of equal opportunity; responsibility and recognition of duties towards the local, national and international community; honesty and integrity; advancement of studies and of scientific research at an international level; equity, impartiality, loyal collaboration and transparency.

It takes on as an essential value well-being in workplaces and in the spaces used for studying.

It adheres to the principles and to the practices of environmental and social sustainability, adopting strategies and behaviour aiming at minimizing its impact on the environment and on natural resources, at increasing social cohesion and at reducing inequalities, at favouring cultural growth and sustainable economic progress' (Ca' Foscari University, 2012a, p. 3).

Sustainability therefore becomes a fundamental value for Ca' Foscari and the sustainable development model represents the target paradigm that inspires choices and actions at all levels, of governance, strategic, organizational and managerial, in an ever-broader process of stakeholder engagement, inside and outside the university's institutional boundaries.

Furthermore, the main aspects of Ca' Foscari's organization with respect to sustainability are presented, with particular reference to the organizational bodies at the top levels, to the main goverance tools significant in the perspective of sustainability (Statute, Strategic Plan and Ethical Code) and to the mapping process of stakeholders, a central element in a goverance vision extended to the various parties with an interest in the organization.

DOI: 10.1057/9781137351937

3.2.1 Sustainability and bodies of governing, managing, control and protection

Since 2009, when the new Rector was elected, Ca' Foscari has made important steps down the path towards sustainability, with two main objectives:

- To create a site of 'sustainable' research, education and work;
- To lead Ca' Foscari to the highest levels of international recognition on the themes of sustainability, coherently with its prestigious status as one of the oldest commercial schools in the world.

The Rector can delegate to Pro-rectors and Representatives chosen among the University professors. A main decision on the perspective of the success of the sustainability strategy was made in 2009, designating, as anticipated previously, a Representative for 'Environmental sustainability and social responsibility of the University'. The Representative is entrusted with overall responsibility for the sustainability themes in the university, which takes effect in the supervision and co-ordination of the substantial series of actions, initiatives and projects in the matter of environmental and social sustainability that Ca' Foscari has planned and will pursue in the future, among which is the writing of the sustainability report, representing a main contact point for the different categories of internal and external stakeholder (Locatelli and Schena, 2011). With this decision to appoint a specific delegate to sustainability, the university's top management intended to promote the high quality of the scientific expertise present in the university on sustainability themes. The figure of the Representative further highlights the strategic importance of sustainability for Ca' Foscari, because of the strong push towards sustainability inside the whole organization and contaminating all processes, starting from the most relevant issues. The benefits at the organizational and procedural level are represented above all by the light procedures, faster and more effective, targeting the strategic objectives of the university – decisions optimizing at the same time effectiveness, costs and sustainability profile.

The Academic Senate, the governing body headed by the Rector, is responsible for the approval of the Ethical Code, once approved by the Board of Governors; moreover, it provides an opinion on the Sustainability Commitments Charter, before the resolution of the Board of Governors.

DOI: 10.1057/9781137351937

The Board of Governors is required to express an opinion on the Ethical Code and on any changes to it, as well as being called to approve the Sustainability Commitments Charter, taking into consideration also the Academic Senate's opinion.

The governing bodies of Ca' Foscari, therefore, according to the Statute, are given precise functions and responsibilities with regard to sustainability, which are also declared in the expression of mandatory opinions and in the approval of sustainability documents, such as the Ethical Code, the Sustainability Commitments Charter and the Sustainability Report, or documents that also contain strategic policies of sustainability, as in the case of the Strategic Plan. In particular, for the Sustainability Commitments Charter, a periodic monitoring process was set up, which involves mainly the Academic Senate and the Board of Governors.

Among the advisory bodies and those of protection, there are the Assembly of Student Representatives, the Council of PhD Students (these two bodies are called on to express their opinions, for the parts under their power of decision, on the Sustainability Commitments Charter and the Ethical Code) and the Student Representative, whose role is to protecting the interests and promote the expectations of two important categories of stakeholder (undergraduate and graduate students). In addition to these two, the university has established a committee whose role is particularly important in the sustainability theme, above all with reference to the social dimension. It is the commitee for equal opportunities, dealing with the improvement of the well-being of employees, that fights against discrimination; it is the committee that 'promotes initiatives for the realization of equal opportunities and the enhancement of the differences between man and woman according to the Italian and European community legislation in force, sees that the principle not to discriminate on gender or sexual orientation is respected and guarantees support to the victims of transgressions and abuse of power. The Commitee also sees that no mobbing is done within the university' (Ca' Foscari University, 2011c, p. 19).

3.2.2 Ca' Foscari's Statute and the Sustainability Commitments Charter

The Statute is one of the cornerstone documents in the governance system, regulating the execution of the activities of the university. Within it,

DOI: 10.1057/9781137351937

among the others, the fundamental principles are pointed out, dealing with internal rules among different bodies, their responsibilities, trying to balance the interests of different stakeholders groups. In the Statute are shown the aspects pertaining to the university organization, with the different bodies that constitute it, respective roles and responsibilities, as well as the organization of the didactics and research structures being represented.

In March 2011, as mentioned previously, the new University Statute (Ca' Foscari University, 2011c), which contains important innovations from a sustainability point of view, became effective. The document represents a further step towards the sustainability goals pursued by Ca' Foscari; the Statute puts the university at the forefront on the sustainability drive, since the Venetian university is the first in Italy to have a Statute aligned with the new directions dictated by the reform of the university system (law 240 of 30 December 2010).

With the new Statute, the process of stakeholder engagement is further strengthened, for example, by the provision of the already mentioned Council of PhD students, and by the inclusion of administrative technical staff representatives in the Academic Senate and in the University Board of Governors.

The centrality assigned to sustainability is made obvious in the targets indicated in the scope of the 'Principles related to the university action' (Article 3), where it is affirmed that 'the University guarantees equal opportunities in research, in education and in work. The University gets an "Ethical Code", a "Code of Conduct for the prevention and the fight against the mobbing phenomenon" and a "Code of Conduct against sexual harassment", aiming at avoiding any form of discrimination internally, direct and indirect, related to gender, to age, to sexual orientation, to ethnic origin, to disability, to religion and to language, every type of conflict of interests and any form of nepotism and favouritism, for the prevention of sexual and moral harassment (mobbing) for the protection of the dignity of female and male workers, female and male students.

The University writes a Sustainability Commitments Charter where the rules and objectives are defined, aiming at minimizing its impact on the environment and on the natural resources, at increasing the social cohesion and at reducing the inequalities internally, at favouring the cultural growth and the sustainable economic growth of the territory.

DOI: 10.1057/9781137351937

It assumes, as a fundamental value, the wellbeing in the study and work sites and prepares prevention strategies to improve the safety and the overall quality of its activities.

It favours, through its advisory bodies and those of proposal, the participation of all its components' (Ca' Foscari University, 2011c, p. 5).

The fact of including in the basic priniciples of the university concepts such as equal opportunities, impact on the environment and on natural resources, social cohesion, inequalities, cultural and sustainable economic growth of the territory, and well-being in study and work sites, makes such aspects rise to fundamental dimensions of the actions of the university, as evidence of a pervading approach, not limited to one-off actions perhaps of a merely solidarity nature or only to communication activities towards the outside. Some of the main documents are mentioned in the principles with which Ca' Foscari controls and manages the fundamental sustainability themes, from the Ethical Code, the Code of Conduct for the prevention of and the fight against the mobbing phenomenon and the Code of Conduct against sexual harassment, to the Sustainability Commitments Charter.

Further space is given to the Ethical Code and to the Sustainability Commitments Charter in the Statute, with articles specifically dedicated to them, where the main aspects of the documents are described.

But it is especially article 53 of the Statute that represents an element of strong innovation for the Statute. The article is dedicated to the Sustainability Commitments Charter, which 'defines the objectives aiming at minimizing the impact of the University on the environment and on the natural resources, at increasing social cohesion and at reducing the inequalities internally, at favouring the cultural growth and the sustainable economic growth of the territory' (Ca' Foscari University, 2011c, p. 34).

The Sustainability Commitments Charter is a volontary tool, not imposed by legislation, as is the case of the Ethicalal Code, and at present, in the national university situation, it has adopted only by the Venetian university. By introducing an article in the Statute expressly dedicated to sustainability, Ca' Foscari effectively states with even more evidence the adoption of the model of sustainable development, basically an irreversibile decision, considering also the complexity and the duration of the approval procedure for the Statute.

The formulation of the article itself and the sequence of the concepts presented imply a precise philosophy, putting the issues concerning the environment in the foreground, talking first of all about minimizing the

DOI: 10.1057/9781137351937

impact of the university on the environment and on natural resources. The initial emphasis on the environmental variable is ascribable to the fact that, in the previous part of the Statute, the focus instead is essentially on the social dimension. It is thus intended, with the wording set out this way, as it were to 'reclaim' the environmental dimension, asserting the significance consistently with the multilevel meaning of the sustainability, which in its ideal acceptance consists in the synergic and balanced integration of the three constituent dimensions. If they are quite commonly agreed on in the profit organizatios, the repercussions on the ecosystem generated by the university institutions are really an innovation for a university, and represents a high level challenge in the case of Ca' Foscari, considering the unique characteristics of the territory/lagoon where the University stands. With Article 53 of the Statute Ca' Foscari finds a balance among the dimensions of sustainability, affiming its adherence to the concept that the three constituent elements integrate in a balanced and synergetic way.

3.2.3 Ca' Foscari's Strategic Plan

The Strategic Plan represents the main document of the course of action of the university. In it, besides a report on the conerstones of the basic orientation (values, mission and vision), the target context is outlined, in terms of opportunities, strengths and organizational and environmental obligations, identifying the long-term path of the institution, described in the strategic objectives, each of which is equipped with specific lines of strategic actions.

In May 2011 Ca' Foscari's Board of Governors approved the new strategic plan 'Towards Ca' Foscari 2018' (Ca' Foscari University, 2012a), where themes connected to sustainability permeate the document and become the university's strategic objective. The first clear reference to sustainability is right at the beginning, in the first lines of the presentation of the Strategic Plan by the Rector, where it is affirmed that 'Ca' Foscari is today more than ever called to react to the changes in the economic and cultural context, but also social and environmental ones, that the crisis, first financial, then productive and in the sustainability perspective, has drammatically quickened' (Ca' Foscari University, 2012a, p. 1). This represents a strong signal by the university in diffusing from the beginning the role of sustainability in its strategic path, defining it as an unavoidable paradigm for all stakeholders and in particular for institutions such

DOI: 10.1057/9781137351937

as universities that are required to carry out an essential task for the development and the growth of the social and economic system.

Sustainability is presented as an opportunity to be seized when it is affirmed that 'the emerging trends lead to achievement of an economy always more sustainable, experiential and global. It is always more focused on the east. The general growing sensitivity towards the theme of the sustainability constitutes an important opportunity to promote Ca' Foscari, given the important research projects on the theme of sustainability developed in the last years by the professors both in the scientific and economic areas' (Ca' Foscari University, 2012a, p. 11).

Such an affirmation is then translated concretely into the identification of the strategic objectives and of the relevant lines indicated for the pursuit of the same objectives. Among the ten main objectives of the university's strategy, for the objective 'Strategically reorganizing the research and educational activities', the strategic line 'Specialization in: economy and management, conservation and cultural productions, International relations, studies on Asia, environment and sustainability', identifies precisely in 'environment and sustainability' as one of the thematic scopes to focus on: 'in a world always more attentive to sustainability, Ca' Foscari was the first University in Italy to establish a University Course in Environmental Science. Environmental protection and reclamation, sustainable development of the territory, energy efficiency and the production of renewable energies are unavoidable objectives today. Ca' Foscari can become a centre of excellence for the interdisciplinary study of the economy, the environment and society, promoting collaboration with other research institutions... in Venice, so that the process of improving environmental performance is also an opportunity for the creation of competitive advantages sustainable in the long term' (Ca' Foscari University, 2012a, p. 11).

But it is above all with Objective 10, 'Assuming a transversal sustainability orientation', that sustainability is recognized as a fundamental target for the acting of the institution, 'transversely' implying a non-sectoral or departmentalized approach, but one that synergetically pervades all the levels, with respect to stakeholders, and that therefore must permeate each and every decision and action.

For the pursuit of this objective the university identifies three lines of action, to be implemented in the three-year period 2012–14 (Ca' Foscari University, 2012a, pp. 52–3):

DOI: 10.1057/9781137351937

- 'strengthening of the sustainabiltiy didactics', recognizing the education activity as a social function, besides the one, perhaps more primary and immediate, of the education of the students for their integration into the world of work. 'The teaching activity must not only be able to prepare the young for the world of work. It must also carry out a social function, it must teach sustainable behaviour, shape new generations of responsible citizens, whatever scope they find themselves making choices in. That's how the didactics, in all its scopes, is called to improve the social fabric, to produce positive externality, to influence the individual sensitivities improving them, making them aware of the need to assume sustainable behaviour, both from the social point of view and the environmental one. Didactics, consequently, of quality also from the social point of view, oriented at supplying the tools not only to excel in the world, but also to make it better.'
- 'development of the sustainablity research', as basic factor, driving where to insert the organization in an effective teaching offer oriented at the sustainability. 'A teaching methodology that teaches to assume sustainable behaviour, that is able to influence positively in the scope of the individual responsibilities it must be supported and, therefore, come from, a research activity intent on the social and environmental aspects. Orienting the research towards such themes, transversal in all the scopes, is the commitment that Ca' Foscari intends to pursue also through the increase of the dedicated research projects. It is important to boost the quality research, the quality research is that which incorporates the sensitivity towards the world that surrounds us inside the objects of study.'
- 'favouring the acquisition of sustainable processes and behaviour', line of action in which the pervading of the approach adopted by the University and the strategic priority assigned to the sustainability is made ovvertly clear, which the management principles, decisional and behavioural processes of the different subjects must follow. While the two previous lines of strategic actions were focused on the main institutional fields of action of the university (didactics and research), this policy presents a broader and more general nature, constitutes a declaration of commitment that goes beyond the primary aim of the University, in full awareness that the model of sustainable development requires an approach that cannot know and reduce boundaries arbitrarily

DOI: 10.1057/9781137351937

> defined by competences legislatively regulated. 'The adhesion to the sustainability policy by whomever studies and works in Ca' Foscari is fundamental. The cultural change already in effect, based on assuming one's own responsibility and on the awareness in introducing daily actions based on sharing the long term objectives, must be encouraged and "embedded" through the optimization of the expertise and resources available, leaning on technology and on innovation as motors able to marry the challenges pointed out by the shortage of the resources and by the precariousness of many natural environments. Decrease in the polluting emissions, cautious management of the water and energy resources, reduction of the waste and increase of the separate waste, decrease of the consumptions, dematerialization, environmental responsibility, attention to social themes, giving value to people and respect for the basic values indicated in the statute and confirmed in the ethic code, must become management principles and elements included in all the decisions, but, overall, they must become part of the "instinctive" behaviour for all of us.'

The description of the strategic lines identified to achieve the objective is assisted by the appointment of the political body, in the specific case of the Representative to the environmental sustainability and social responsibility, as well as by the performance measurement system, that is based on sustainability indicators according to which the results achieved are evaluated.

In addition to Objective 10 and the contents presented for Objective 1, in general the majority of the objectives outlined in the Strategic Plan make reference to multi-stakeholder goverance (Locatelli and Schena, 2011), one of the basic requirements for the acceptance and implementation of sustainability logics. The importance given to the involvement of the internal and external stakeholders (Shriberg, 2002) could be derived by the results and the procedures performed by the management.(Savan and Sider, 2003) is understood.

3.2.4 The Ethical Code

The approval of the Ethical Code is among the first concrete affirmations of the adherence to the sustainability paradigm by Ca' Foscari, considering the fact that the Venetian university anticipated by two years the text of the legislative provisions (law 240 of 30 January 2010) that made such a document obligatory for university institutions.

DOI: 10.1057/9781137351937

In the preamble to the Code, at point 1, comma 1, it is declared that Ca' Foscari University of Venice, 'aware of the important social and educational function carried out by the European university, reflects the values that are historically at the heart of scientific research, of education and of the other numerous university activities and shapes its work according to them in order to favour the development and the spreading of knowledge [and] the creation of an environment marked by dialogue and by proper interpersonal relationships, by openness and by exchanges with the international scientific community, by the teaching of basic values stated and gaurenteed by the Italian Constitution, in the Charter of Nizza and in the other statements of basic human rights signed by Italy and other european countries' (Ca' Foscari University, 2008, p. 1). In comma 2 the cornerstone target values for university institutions are stated: that each member of the university must respect, protect and promote: '(a) human dignity; (b) refusal of every unfair discrimination and enhancement both of the merit and of the individual and cultural diversity; (c) respect of freedom and of basic rights; (d) enhancement of equalities; (e) responsibility and recognition-fulfilment of duties towards the community; (f) honesty, integrity and professionality; (g) incentivating studying and scientific research; (h) equity, impartiality, loyal collaboration and transparency' (Ca' Foscari University, 2008, p. 2).

Ca' Foscari University has started the procedure to endorse a new Ethical Code, since the version currently in force dates back to 2008. It could not refer to the important principles introduced by the sustainability strategy, showing how its strategy had strengthened during the years following 2008 and that led to inverventions in other areas, mandatory and voluntary, such as the Statute, the Strategic plan and the Sustainability Commitments Charter, along with actions carried out on the organizational side.

3.2.5 The mapping of stakeholders

The mapping of stakeholders represents an unavoidable passage, not without elements of criticality, for any organization that intends to activate a course of growth prompted by the logics of sustainability, including the university institution. The process of mapping stakeholders is necessarily a personalized course and defined for its specific purpose, presenting in itself a high level of complexity both in diachronic terms,

DOI: 10.1057/9781137351937

and synchronic. This complexity is intensified if the university follows consistently the sustainability principles.

Ca' Foscari continued with the awareness that this process does not end with the mere matter of identifying stakeholders at a formal level, but has a substantial impact, due to the consideration of the expectations of the stakeholders, changes that must find suitable answers in the decisional processes and in the actions introduced.

The mapping of stakeholders for Ca' Foscari has been realized by a strong process that required an elevated commitment, it is deeply described in both the yearly sustainability reports made so far, in the first section, dedicated to the institutional perspective.

The result of the identification process of the stakeholders turned into a particularily complex mapping. Various factors contributed to generating this result, substantially ascribable to the fact that a university institution like Ca' Foscari presents hybrid characteristics, belonging at the same time to the sphere of the civil service – considering, for example, the importance of public, state or European community financing and its aim to educate and to diffuse culture – and to the sphere of private organizations, if the autonomy granted to the university by the recent legislation is considered (Sammalisto and Arvidsson, 2005).

This leads the university to interact with a variety of subjects, with characteristics of heterogeneity in terms of interests and needs to be satisfied, this diversity being due to the aim and the objectives of their own interests. The process of mapping stakeholders was started in 2010 and became a successful action, through the benchmarking realized considering both profit organizations and foreign universities (considering the absence of university sustainability reports at the national level). The benchmarking was carried out consulting the sustainability report published on line (2009) with the aim of identifying the selection criteria and the methodologies adopted for the mapping of the stakeholders, the analysis of the respective needs and the definition of the objectives. In particular, the Sustainability Balance of Enel S.p.a. and the Sustainability Report of the Swedish University of Gothenburg reports were found to be helpful and inspiring for the stakeholders. The strength of these two reports were recognized in the following ways: high level of clarity, the considerable ability to summarize relevant items, the materiality of the information included, together with the outstanding stakeholders' engagement process.

DOI: 10.1057/9781137351937

This first phase enabled the identification of a kind of objective model. On the provisions of this result, not having an actual database, the Office compiled a preliminary list of the subjects (institution and organizations) that influence Ca' Foscari and proceeded from the analysis of the current actions that generate impacts on stakeholders, with a particular view on sustainability themes. At the procedural level, the greatest difficulty was the absence of a clear and explicit mapping of the actions, read though the lens of the respective needs that they were destined to satisfy. This entailed an analysis of all the actions introduced by Ca' Foscari in the year of reference and a sustainability interpretation of them. Among the numerous information sources that the University used to carry out these complex analyses were:

- Ca' Foscari's Statute;
- The University Strategic Plan;
- The educational offer and the research projects;
- The reporting plan of Ca' Foscari University Venice;
- The Ca' Foscari institutional site (www.unive.it);
- The Infoscari site (www.unive.it/infoscari);
- The Rector's blog, a web page with free access where the Rector informs, by means of a blog, the Ca' Foscari community about events and news considered of particular importance and interest;
- The Sustainability Commitments Charter;
- The Sustainable Ca' Foscari site;
- The site www.fondazionecafoscari.it;
- The Sports Centre site www.cusvenezia.it;
- The feedback of didactics and of general services, feedback based on specific surveys and questionnaire;
- The site www.almalaurea.it, a national database that monitors the careers of graduates.

This analysis was conducted with reference to the previsions of the Statute, of the Strategic Plan and of the Rector's blog; so the structural attention of the university towards the stakeholders and the involvement of the strategic leaders in the process were confirmed.

In accordance with the indications contained in the Guidelines of the GRI (GRI, 2006), the stakeholders identified were initially divided by typology, between internal and external; for each stakeholder category different homogeneous sub-categories were identified, according to the characteristics of the subjects and their respective needs/expectations.

DOI: 10.1057/9781137351937

Among the internal stakeholders the following categories were identified:

- Students, subdivided into the following sub-categories:
  - Bachelor degree students,
  - masters,
  - graduate students,
  - students' families;
- alumni;
- staff, subdivided into the following sub-categories:
  - adminstrative technical staff,
  - teaching staff,
  - researchers,
  - grant winners,
  - others.

The external stakeholders were subdivided as follows:

- community, subdivided into the following sub-categories:
  - local community/citizenship,
  - businesses,
  - haulage contractors;
- suppliers;
- environment;
- partners;
- institutions, subdivided into the following sub-categories:
  - foreign universities,
  - Italian universities,
  - research organizations,
  - Ministry of Education, of the University and of Research (MIUR),
  - Ministry of Economy and of Finance,
  - Ministry of the Environment and the Protection of the Territory and of the Sea (MATTM),
  - European Community and Veneto Region,
  - Venice Commune and Province of Venice,
  - CRUI (Conference of the Rectors of the Italian Universities),
  - Regional Agency for the Prevention and Protection of the Environment of Veneto (ARPAV).

Once the identification of the categories and sub-categories of stakeholder had been completed, the definition of the lines of action

DOI: 10.1057/9781137351937

TABLE 3.3 *Table of stakeholders – extract: student category*

Expectations	Positive externality – Ca' Foscari's lines of action
Receiving education without discrimination of any nature	Periodic reporting of the evaluation of didactics and services; Increasing student satisfaction through proper actions
Obtaining education suited to the interests of the subject	Periodic reporting of the evaluation of didactics and services; Increasing the student satisfaction through proper actions
Obtaining education useful for the world of work	Final reporting of the final-year students (Almalaurea). More graduates of Ca' Foscari find jobs within one year and within three years of their degree than the national average. Periodic meetings with external stakeholders. Meetings between students and employers (e.g. Finance Day, Career Day) organized by the Internship and Placement service
Living away from home	Housing Office Service, Lodgings and Canteens. University lodgings for students off the premises and assigned through competition (merit and income); ESU Service for lodgings and canteens. The lodgings/halls of residence organized by Ca' Foscari in structures positioned in various points of the city can be equated to small campuses to favour exchange and aggregation among students, even for foreign students. Defined the project for the new student residence in San Giobbe (sleeping 300) the making of the new students' residence was started in Campo dei Gesuiti (sleeping 200), in collaboration with IUAV

Source: University Ca' Foscari, 2013d.

undertaken by the university with respect to each of these was proceeded with. The information sources used for the mapping were reviewed again and considered in a multi-stakeholder perspective, with the intention of identifying the repercussion that each action could cause to the subjects indicated in the mapping. The identification of the needs and expectations of the different categories of a subject was made by a deductive approach, reading and classifying the analysis of the relationship defined between stakeholders and actions with the need/expectation that the action intended to satisfy for the categories of subjects concerned

The mapping of the stakeholders carried out by Ca' Foscari was presented in three different manners, depending on the medium used to transmit the information:

DOI: 10.1057/9781137351937

- On the Sustainable Ca' Foscari site the mapping was presented in full, by means of a navigable table subdivided per stakeholder. In this way each user, clicking on the desired sub-category, could visualize the detail containing the expectations and the actions undertaken by the university; internal stakeholders who had access to the restricted area of the site could also examine the information sources;
- In the Sustainability Report (complete version), in line with what had been done by other universities, for example in the Sustanability Report 2009 of the University of Gothenburg, the analysis was reported with a summary table that indicated the main categories of stakeholder and Ca' Foscari's actions, without disclosing the expectations or needs of each stakeholder. In the short version of the Sustainability Report the mapping of stakeholders was not reported, for reasons of space.

In Table 3.3 an extract of the integral mapping presented on the Sustainable Ca' Foscari site is presented, with reference to the category 'Bacheor Degree students'.

3.3 The sustainability tools

The strategic-organizational approach adopted by Ca' Foscari emerges also in a variety of tools introduced right from the first steps along the path towards sustainability, in which this paradigm has evidence of implementation and visibility towards the outside.

They are mixed tools, with distinct significances, some more wide-ranging, which capture sustainability as a whole, some more targeted and concentrated on specific aspects, but always founded within an organic framework.

It is gathered therefore also from the combination of the tools activated by the Venetian university and the integration of the approach; this approach emanates from a unitary standpoint and enables the development of important synergies, practically impossibile to find in an approach of a technical-technological nature. The technical-technological approach is instead expressed in one-off initiatives, perhaps keeping with expectations and sensitivities of the moment, not ascribable to a common orientation.

In the following paragraphs, the sustainability tools adopted by Ca' Foscari are described, with particular attention given to the Sustainability

DOI: 10.1057/9781137351937

Commitments Charter, the Carbon Management Project and the Sustainability Report, before we dwell on more targeted projects and initiatives, but all the same fundamental to create shared awareness of the sustainability themes among all stakeholders.

3.3.1 The Sustainability Commitments Charter

On 8 July 2010 Ca' Foscari's Academic Senate approved the first Sustainability Commitments Charter, a tool that, as decreed by the Statute, 'defines the objectives aiming at minimizing the impact of the University on the environment and on the natural resources, at increasing the social cohesion and at reducing the inequalities on the inside, at favouring the cultural growth and the sustainable economic growth of the territory' (Ca' Foscari University, 2011c, p. 34).

The Venetian university is the first Italian university to have such a tool, a direct result of the sustainability policy of the university, characterized by a multi-value semantics respect for the sustainability themes.

With the publication of the Commitments Charter 2010–12 Ca' Foscari for the first time presented objectives, commitments, indicators, targets and managerial control for a three-year period (2010–12), from the standpoint of economic, social and environmental sustainability.

The Sustainability Commitments Charter was organized in ten areas of intervention: the governance/institutional area, policies for students, policies for staff, the supply chain, energy, water, materials, waste, mobility and innovation.

The following are indicated for each:

- Strategic objectives and operational objectives;
- Commitments towards stakeholders;
- Indicators of results/impacts, for the evaluation of the actions and outputs;
- Time targets (specific for each of the three years of reference);
- The organizational unit responsible for the pursuit of the objectives and for the realization of the commitments.

Ca' Foscari clearly expressed the strategic and operational objectives concerning the tool itself, that there was an intention to pursue, through a Commitments Charter, objectives that had been communicated to stakeholders in the course of a meeting ('Sustainability at Ca' Foscari: towards a sustainable university'), held in March 2011. The

DOI: 10.1057/9781137351937

choice to officially present the Charter to stakeholders confirms the participatory approach adopted by the university as to sustainability. In particular, at the strategic level, with the Commitments Charter Ca' Foscari intended:

- To initiate a process of explicit and shared change;
- To pursue strategic objectives in a multi-stakeholder logic;
- To improve the image of the university and the satisfaction of its stakeholders.

At the operational level, with the Commitments Charter the university intended to activate a process of governance of the sustainability themes, oriented at:

- Modelling the structure of the operating systems according to the ten areas identified;
- Managing in a systematic manner, with specific plans of action in the short and medium term, the indivual projects announced by the relevant organizational divisions/units;
- Activating a process of periodic monitoring of the state of progress of the projects and a yearly review of the Charter.

In the different editions of the Commitments Charter there was a succession of interventions aiming at perfecting the tool and the process to which it relates, from the integration of indicators of results and of impact used for the evaluation of the achievement of the objective, to the direction of the operating units involved, in collaboration with the unit responsible for the managerial controlo. The latest edition of the Commitments Charter, concerning the period 2013–15 (Ca' Foscari University, 2012b), approved by the Board of Governors of the University on 21 December 2012, compared with the versions that preceded it, contained a smaller number of objectives, but with a superior level of detail, a decision that has allowed for the clear expression of more customized targets.

Of particular importance are the characteristics of the process activated by the university as to the Commitments Charter, consistently with the strategic and operational objectives stated previously, from its construction to its monitoring and review/updating.

The construction of the Commitments Charter was the result of a process co-ordinated by the Representative of the Rector for environmental

DOI: 10.1057/9781137351937

sustainability and social responsibility, with the collaboration of the Special Processes and Projects Office and the proactive involvement of the different managements and organizational units of the university. In the identification of objectives and commitments it is not the organization (organizational structure and organizational mechanisms) but rather the orientation towards the stakeholders that prevails; therefore much evidence is given to the planned results, setting aside the structure and the intra-organizational boundaries, from a transversality point of view. This result-oriented strategy drives the promotion of a collaborative approach among the various organizational units, to overcome the rigidity connected to silos/departmentalized visions, that often involve lower levels of effectiveness and efficiency.

The state of advancement of the actions undertaken and the achievement of the objectives stated in the Charter are subject to systematic monitoring at monthly or quarterly intervals by the governing bodies of the university and to yearly reviews, which can be achieved both in a reformulation of the objectives and in the introduction of new actions. Ca' Foscari furthermore reports to stakeholders every six months the state of advancement of the commitments stated in the Charter, thanks to close monitoring, in collaboration with the responsible organizational units involved in the execution of the actions.

By way of an example, in Charter 2013–15 the area 'Policies for the students' is organized as described below.

For the pursuit of the strategic objective 'promoting the satisfaction of the students' two operational objectives are identified:

a) To support the involvement and the participation of students in the sustainability themes;
b) To support the right to education and to increase the efficiency and the effectiveness of the services to the students.

Each operational objective is linked to one or more 'commitments towards stakeholders', which represent the modalities through which the university intends to pursue the objective, monitored through indicators of results/impact and relevant yearly targets, in addition to information on the organizational unit responsible.

The details of the two above-mentioned operational objectives (called objective a) and objective b)) are given below, omitting the information on the organizational unit.

DOI: 10.1057/9781137351937

TABLE 3.4 *Sustainability Commitments Charter 2013–15: objective 'a'*

Operational objective	Commitment to stakeholders	Indicators of results/ impact	Target		
			2013	**2014**	**2015**
Support the involvement and the participation of the students in the sustainability themes	2.1 To involve the student representatives bodies in targeted meetings and, in particular, obtain the support of the committee that evaluates the financing projects of the initiatives self-managed	A) Number of students participating in the activities; B) Number of targeted meetings	Monitoring Number of students participating and number of targeted meetings	Continuous	Continuous
	2.2 To found awards for theatre, musical and cinematographic projects dedicated to sustainability and having an educational nature	A) Number of awards; B) Number of participants	A) Start projects; B) Record subjects involved	A) Execution projects started and start of new projects; B) Increase subjects involved	A) Execution and strengthening of projects started; B) Increase subjects involved

Source: Ca' Foscari University, 2012b, p. 3.

DOI: 10.1057/9781137351937

DOI: 10.1057/9781137351937

TABLE 3.5 *Sustainability Commitments Charter 2013–15: objective 'b'*

Operational objective	Commitment to stakeholders	Indicators of results/impact	Target		
			2013	2014	2015
Support the right to education; increase the efficiency and the effectiveness of the services to the students	2.3 To extend the library services into evening hours	Installation of self-loaning devices to obtain books without the assistance of employees in the libraries with extended opening hours	No. 2	No. 2	
	2.4 To enter into a special agreement with the local health department (ULSS 12 Venice) to assure health care for students 'off premises' both in the commune of residence and in the territory of the ULSS 12	A) Number of students off premises with an agreement	Start of contacts	Stipulation of the agreement	Arranged and effective agreement
	2.5 To extend payable services by multi-purpose card (Paypass)	On/off	Four-monthly monitoring	Four-monthly monitoring	
	2.6 Application of procedure for the correct management of the complaints and grievances process	A) % reduction complaints and grievances	Four-monthly monitoring	Four-monthly monitoring	Four-monthly monitoring

Source: Ca' Foscari University, 2012b, p. 3.

3.3.2 Carbon Management Project

On 23rd July 2010 Ca' Foscari, in the scope of a Memorandum of Understanding with the Ministry of the Environment and of the Protection of the Territory and of the Sea (MATTM), entered into an agreement for the realization of a pilot project for carbon management.

There were essentially two outputs of the joint project:

- Planning and implementation of an effective model for the calculation of the carbon footprint, with the consequent proposal of corrective measures targeted at reducing the emissions registered;
- Publication of Guidelines for Carbon Management in Italian Universities, a tool that aims to represent a target for the other Italian universities that intend to adopt a system for the management of the CO_2 emissions, providing support in the phase of identification of the methodologies and of the technologies of reference.

At the strategic level, the Carbon Management Project intends to pursue aims that pertain to various dimensions of sustainability (and therefore goes beyond the law prescriptions), while compliance with the law is pursued to try to improve the quality of regulation compliance. Through the project, the university aims to achieve positive results both on the economic front, thanks to the rationalization and reduction of the costs that result from the implementation of the mitigation measures of the negative impacts linked to the carbon footprint, and on the environmental front, the pursuit of a more eco-friendly positioning of the university, all this also thanks to greater awareness among stakeholders of the themes in question, with a consequent diffusion of a sustainable life style.

At the operational level, as mentioned previously, there are various activities that must be implemented for an effective and efficient control of all the aspects pertaining to the carbon footprint: from the implementation of an information management system for energy needs, the systematic recording of energy consumption, and the execution of the mitigation measure of the environmental impact registered to the modernization of the infrastructure, in order to improve the level of energy efficiency. In the perspective of constant improvement in the management of energy consumption, the university also developed an incentive plan for the more 'virtuous' structures, or rather those with the best energy performance, thanks to a sensible use of energy resources.

DOI: 10.1057/9781137351937

This required accurate monitoring of the energy consumption of the various structures, according to the relevant half-yearly reports, made by comparing data derived from management accountancy with analytical information provided by the tool of remote control.

The collaboration with the MATTM continued thanks to a new agreement called 'Addendum' (entered into on 14 July 2011 and approved by the Board of Governors on 7 October 2011), this too with a double objective:

- The attainment of the ISO 14064–3:2006 certification for the proposed methodology for the calculation of the carbon footprint;
- The definition and implementation of innovative actions aimed at spreading the principles of sustainable behaviour, promoting the choice of a responsible life style (within these falls the 'CO_2 calculator', which will be discussed later on).

The realization of the Carbon Management Project, with the quantification, reporting and certification activities of the CO_2 emissions, together with further development lines provided for by the 'addendum' project, required a strong commitment by the university, involving the whole staff.

The Carbon Management Project activated by Ca' Foscari substantially followed the approach indicated by the Deming Cycle (Deming, 1950), based on the sequence 'Plan-Do-Check-Act', developed according to the logics illustrated in Chapter 2.

Some of the peculiar aspects that characterized the path of the Venetian university in the realization of the project are presented below.

In general, Ca' Foscari met its own energy demands using the national electrical grid for lighting and air-conditioning, while heat was generated by boilers on site, powered by natural gas supplied by the local grid distributor.

The calculation of the greenhouse gas emissions is essential to provide a baseline that can help to manage the development and the implementation of future strategies concerning the reduction of greenhouse gas emissions, and also to monitor progress towards 'carbon neutrality', a long-term objective of the university.

There were four phases in Ca' Foscari's calculation of its carbon footprint (Ministry of the Environment and of the Protection of the Territory and of the Sea and Ca' Foscari University, 2012):

1. Data collection;
2. Definition of the organizational and operational boundaries, of the reference year and of the functional unit;

DOI: 10.1057/9781137351937

3 Analysis of the data, calculation model and development of the software tool;
4 Presentation and discussion of the results obtained.

The collection of the primary data was carried out by a technical group, comprising representatives of the various administrative divisions of the university, under the supervision of the Carbon Management Team (CMT), itself made up of representatives of the university and of the MATTM. Using relevant thematic questionnaires and surveys, data were collected for the period 2007–10, with the exception of the data on staff travels, which were collected only for the period 2009–10. The data regarding the mobility of students and staff were collected via an online questionnaire and then compared to the total number of students and employees. The quality of the data was evaluated according to precise criteria and classified into two categories: satisfactory or unsatisfactory (for the use of the unsatisfactory ones, a process of estimation was made, methodologically defined and disclosed).

With regard to the second phase (definition of the organizational and operational boundaries, of the year of reference and of the functional unit), 2009 was chosen as the base year, this being the period with the most complete data. The quantity of greenhouse gas emissions was analysed both for 2009 and for 2010, making some comparisons for the CO_2 emissions specific of the Scope 1 and of the Scope 2, since the data for the period 2007–10 were available.

In defining the organizational boundaries it was chosen to use the control approach: according to this methodology, 8 of the 36 settlements were excluded from the analysis since the university does not directly control their management. The same approach was used in the case of the canteens.

The sources of emissions used for the analysis of the Ca' Foscari's carbon footprint are coherent with the indications of the GHG Protocol of WRI/WBCSD and were examined in an in-depth manner (for importance, completeness, consistency and precision).

Some sources of emissions were not considered in the scope of the present study due to the lack of data or to the fact that the university does not have financial or operational control over some activities. The buildings where the university is a passive subject of a tenancy contract were not included either.

DOI: 10.1057/9781137351937

TABLE 3.6 *Sources of emissions for Ca' Foscari University*

Emission scope	GHG emission source
Scope 1	Combustion in the heating, powered by natural gas
	Refrigerant leaks [R134a+R407c+R410c] from the HVAC systems
	Fuel consumption by vehicles owned by the university
Scope 2	Electricity purchased
Scope 3	**Sources of upstream emissions**
	Goods purchased (paper, books, hardware, toner, furniture and fittings)
	Fugitive emissions from the extraction, transportation and distribution of fuels used for the energy needs of the university
	T&D leaks of electricity in the national grid
	Sources of downstream emissions
	Management of waste
	Staff travels
	Mobility of staff
	Mobility of students
	Trips by international students
	Trips by national students

Source: Ministry of the Environment and of the Protection of the Territory and of the Sea and Ca' Foscari University, 2012, p. 115.

The final decision on the emissions included in the analysis, as stated above, was made in compliance with the requirements requested by the GHG Protocol of WRI/WBCSD: the minimum requirement entails the obligatory reference of the emissions to the Scopes 1 and 2, and a level of the emissions of the Scope 3 equal to at least 80% of their maximum.

Apart from the mobility of students and staff, no activity carried out by the students or the staff outside the university boundaries were taken into consideration. Not even some goods purchased falling within the upstream emissions (for example, paper napkins, chemical cleaning products) were included. Furthermore, the fuel used for emergency electric generators was not taken into consideration, given the limited incidence of the emissions, due mainly to periodic maintenance. Emissions related to the purchase of laboratory materials and to the handling of toxic-noxious waste were calculated, but they were not presented in the study under discussion, since the total emissions amounted to less than one tonne of CO_2 equivalent.

The next step in the computation of the carbon footprint was the preparation of a calculation tool, a model developed by using a Excel sheet that contained all the sources of the emissions. One of the sheets of this tool is shown in the Figure 3.3.

DOI: 10.1057/9781137351937

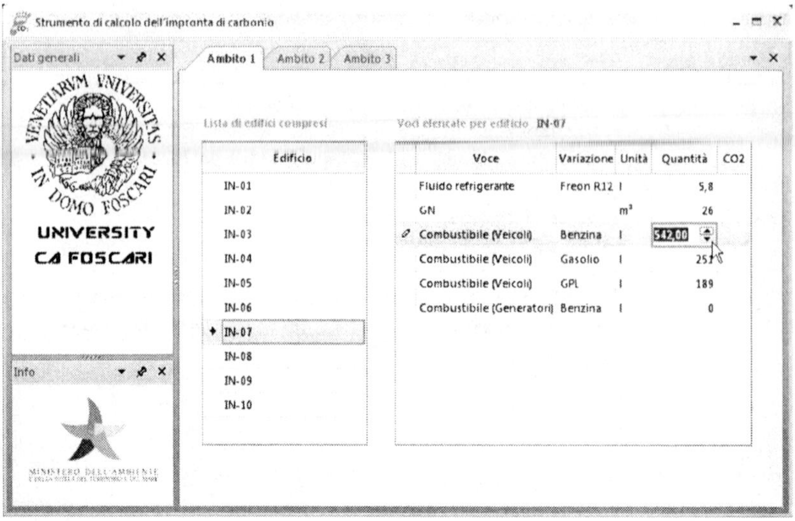

FIGURE 3.3 *User-friendly tool for the calculation of the University carbon footprint*
Source: Ministry of the Environment and of the Protection of the Territory and of the Sea and Ca' Foscari University, 2012, p. 122.

The result of the analysis carried out brought to light the fact that the carbon footprint of Ca' Foscari University (including the complete life cycle of the university services) amounted to 12,414 tonnes CO_2 equivalent for the year 2009 and 12,568 tonnes CO_2 equivalent for the year 2010. A summary of the results, subdivided by individual scope, is presented in the Table 3.7.

With regard to the functional unit, the number of students was identified as the most important driver and therefore also the value of the greenhouse gas emissions per student was calculated, in addition to the one per m^2 (Table 3.8).

According to the actual results, guidelines for identifying the measures for emission reduction were defined, focused on the following aspects:

- Reduction of the building emissions, one of the main sources of greenhouse gas emissions, according to evidence revealed by the calculation of the carbon footprint; these figures have also been audited and the energy audit gave as an outcome the definition of the monitoring systems and plans of action for the activation of mitigation measures of the impact;

DOI: 10.1057/9781137351937

TABLE 3.7 *Ca' Foscari University's carbon footprint – recapitulatory table*

Emissions	Units	Greenhouse gas emissions 2009	Greenhouse gas emissions 2010
Scope 1: Direct emissions of the university			
Consumption of natural gas	tCO_2eq	1,487	1,937
Refrigerant leaks	tCO_2eq	137	137
Fuel consumption – vehicles belonging to the university	tCO_2eq	17	12
Scope 2: Indirect emissions from the use of purchased electric energy, heating and cooling			
Consumption of electric energy	tCO_2eq	3,619	3,560
Scope 3: Indirect upstream emissions			
Input materials	tCO_2eq	374	276
T&D leaks	tCO_2eq	265	228
Fugitive natural gas	tCO_2eq	118	154
Scope 3: Indirect downstream emissions			
Handling of waste	tCO_2eq	211	174
Missions	tCO_2eq	192	218
Mobility of the staff	tCO_2eq	456	456
Mobility of the students	tCO_2eq	4,499	4,329
International students	tCO_2eq	1,010	1,060
Italian students residing outside commune	tCO_2eq	28	26
TOTAL: **Scope 1+ Scope 2+ Scope 3**	tCO_2eq	**12,414**	**12,568**

Source: Ministry of the Environment and of the Protection of the Territory and of the Sea and Ca' Foscari University, 2012, pp. 122–3.

TABLE 3.8 *Greenhouse gas emissions per functional unit in the case of Ca' Foscari University*

Functional unit	Greenhouse gas emissions [$kgCO_2eq$] 2009	Greenhouse gas emissions [$kgCO_2eq$] 2010
Student	574	589
Mq	159	157

Source: Ministry of the Environment and of the Protection of the Territory and of the Sea and Ca' Foscari University, 2012, p. 124.

- Management of mobility; the most significant contributor of greenhouse gas emissions of the university, comes from the commuting and the mobility of students and employees. To reduce such emissions Ca' Foscari has set out policies, commiting itself to promote systems of sustainable transportation;

DOI: 10.1057/9781137351937

- Management of waste, with Ca' Foscari's first essential step represented by setting up a system of separate collection and recycling of waste extended to all the premises, to raise awareness among students and employees, accustoming them to ask themselves what type of waste they are producing and what could be the ways not to generate that type of waste (the Ra.Di.Ca project fits exactly into this context);
- Management of purchases and contracts, which impinges significantly on the emissions coming from the actions of external businesses that work for the university, intervening in the specific techniques contained in supply procedures and reviewed in terms of sustainability (Green Public Procurement);
- Definition of a plan of action and monitoring, targeted at effectively managing the building equipment management system, within a pre-arranged and coherent path (road map) that involves all staff but also the final users, with the common objective of implementing a reduction in emissions over time and at the same time a curb on energy consumption.

Finally, in a strategic-organizational approach the reporting activity is essential, both internal, and external, to circulate the policy choice to undertake a management project of energy consumption and of emissions of carbon dioxide.

In this perspective, Ca' Foscari has activated both internal reporting initiatives, for the awareness, the sharing and the involvement of stakeholders such as students, professors and technical and administrative staff, families, and external reporting actions, essential for the university's reputation and the success of the management of the Carbon Management Project, which constitutes an obvious merit for the social reputation of the university in the relationships with its external stakeholders.

Within the Carbon Management Project, an initiative of high importance was the planning and implementation of the already cited carbon footprint calculator, considered a tool of extreme importance above all to create awareness, a tool generating a strong educational impact. Each daily action involves, in different ways, emissions of carbon dioxide: through the calculator the weight in CO_2 of the daily activities can be measured, suggesting how to make the life style more sustainable and therefore of smaller negative environmental impact. The importance of

DOI: 10.1057/9781137351937

this tool is understood from several sides. As regards the calculation of Ca' Foscari's carbon footprint, the tool turns out to be crucial in order to collect data for the quantification of the emissions concerning Scope 3 and constitutes the main tool for students to use. In broader terms, the use of the Carbon Footprint Calculator and its effectiveness in the engagement of students will work in the long term, aiming at achieving a positive externality from from an educational point of view. As already pointed out, above all in one of the ten objectives of Ca' Foscari's Strategic Plan (Ca' Foscari University, 2012a), 'to take on a transversal orientation of sustainability', the social role of didactics is recognized, targeted not only at shaping the young for their entrance into the world of work, but more broadly to bring up responsible citizens, who will adopt sustainable behaviour and in their turn be actors in the spreading of sustainability principles. The aim is to activate a virtuous course through tools such as the Carbon Footprint Calculator, which will arouse in students, to start with, greater awareness of the impacts directly generated by their behaviour and, through this action of awareness, generate positive repercussions that reproduce over time in the social fabric.

The Carbon Footprint Calculator is a user-friendly tool, accessible online. It is an interactive tool available to members of the Ca' Foscari community (that is those who can access the restricted area, with access credentials), which allows them to calculate their own environmental impact in terms of CO_2 equivalent, in order to develop awareness of the importance of the sustainability themes, helping to spread sustainable behaviour.

In order to use the tool, the user must therefore enter the restricted area and fill in a questionnaire (that takes about ten minutes) developed by Ca' Foscari with the technicians of the Ministry of the Environment, of the Protection of the Territory and of the Sea. The questions included pertain to the following areas:

- Waste, with the typology of separated waste;
- Transportation, with the average time spent on the means of transportation identified;
- Food, from preferred foods to foods eaten most often of various categories (raw and cooked meat, milk and dairy products, fruit, bread, beverages, etc.);
- Energy, from the use of energy-saving lighting to the temperature usually kept in home during the winter months, to the use of

DOI: 10.1057/9781137351937

differentiated water flows in the toilets, to the yearly energy consumption);

- Purchases, investigating, for example, the frequency of purchase of various products (items of clothing, electronic devices, etc.), the types of bag most frequently used for the purchases, the number of shops per week;
- Study, from the principal place of study to the equipment used in studying, to the number of books purchased in a year.

Once (s)he has completed the questionnaire, the user will know which among the daily activities carried out involves the most emissions of carbon dioxide and chooses a strategy to reduce the impact, fixing targets. The data introduced are stored and the user can embark on a programme of reduction of their emissions and re-do the questionnaire at a later time to appraise the improvements achieved. In addition to allowing users to quantify their own carbon footprint and therefore to know how much they 'weigh' on the environment, the tool makes it possible to identify the area for reduction and conservation of the environmental footprint (such as, purchase, study etc.) with suggestions for the adoption of a life style that generates fewer negative repercussions on the ecosystem in terms of polluting emissions.

FIGURE 3.4 *Carbon footprint calculator: home page*

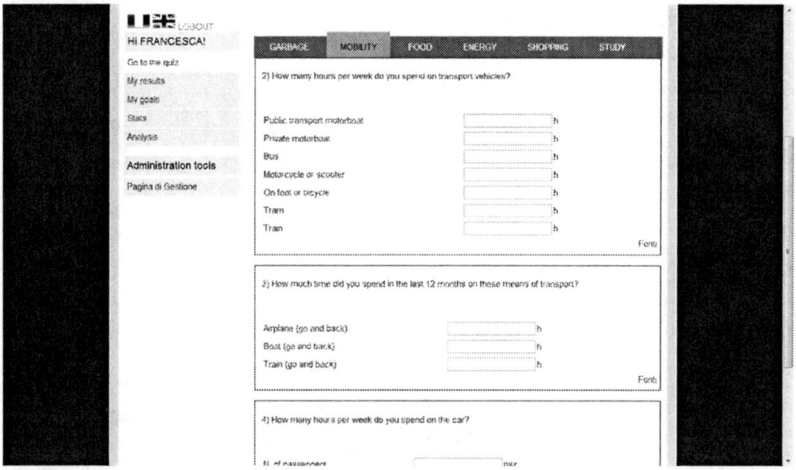

FIGURE 3.5 *Carbon footprint calculator: questionnaire*

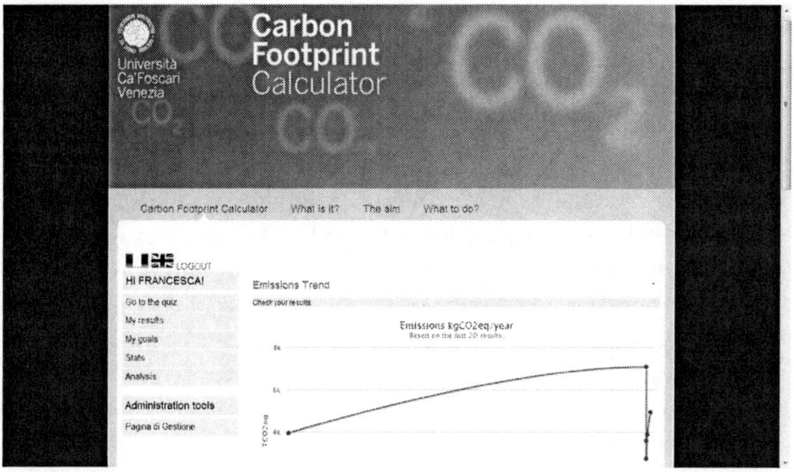

FIGURE 3.6 *Carbon footprint calculator: results (emissions trend)*

3.3.3 Sustainability Report

Ca' Foscari's first Sustainability Report represents a fundamental step on the path towards sustainability, which the university has intended to place as a step within a broader process of management of sustainability. This process finds a critical time in the stakeholders' engagement

DOI: 10.1057/9781137351937

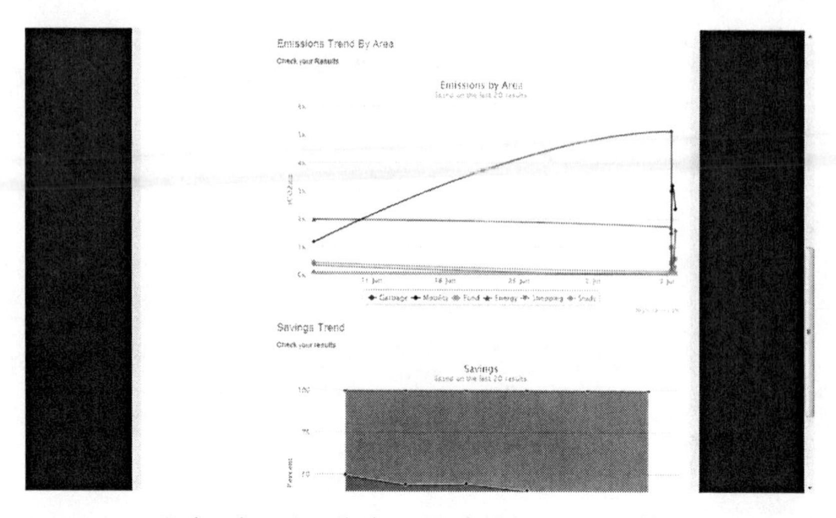

FIGURE 3.7 *Carbon footprint calculator: results (emissions trend by area; savings trend)*

realized through the report, consistently with the organizational strategic approach Ca' Foscari has adopted. The Sustainability Report, therefore, is not a tool of mere reporting value, but constitutes the output of a process on sustainability issues that performs different and complementary functions. In addition to its informative function, as an opportunity to communicate to the outside world the university's work in terms of social, environmental and economic impacts, for Ca' Foscari the Sustainability Report is a management tool, at the same time an opportunity and driving factor to further develop the integration of the sustainability aspects within the strategy and the management systems, and an essential tool with the function of public relations, a tool for activating and continuously fuelling a structured dialogue with stakeholders.

Ca' Foscari's first Sustainability Report (Ca' Foscari University, 2011b) was officially presented on 8 July 2011 by the Rector during an open meeting with all stakeholders.

The process that led the university to this result, beginning with the collecting and subsequent analysis of data, lasted about seven months, from December 2010 to June 2011.

Overall the writing of the report is an output, the fruit of teamwork that sees the whole structure involved, with different modalities and degrees of intensity, with the co-ordination of the Representative and the support of the Special Processes and Projects Office, which had the task of collecting and systematizing the information and of writing the text of the document.

In the first place, Ca' Foscari's objective was to make the strategic-organizational approach emerge from the report, defining the strategic context, the commitment and the sustainability policy of the university, and outlining and strengthening the core elements.

Another important decision was the choice of the principles and of the guidelines to adopt in the writing of the report (framework), which would determine the type of information to collect and the indicators to construct. After a phase of comparison among the various standards proposed at the national and international level, Ca' Foscari decided to adopt as a main reference the Global Reporting Initiative (GRI) model, also in consideration of the high circulation of that standard, which gives greater scope for comparisons among the various organizations and reference periods. It being their first experience of writing such a report, the university's work group decided to make reference to the principles indicated by the GRI as guides in the choice of the contents and of the methods for the evaluation of the information in each theme area, leaving to the subsequent editions improvements and higher levels of detail. Besides the GRI Guidelines, Ca' Foscari adopted methodologies indicated by the CAF University (CRUI Foundation, 2010) and the information coming from the drafting of the self-assessment report that the Venetian university wrote since its participation at the CAF methodology (CRUI Foundation, 2012). In the subsequent edition of the report, in 2011 (Ca' Foscari University, 2012c), the university followed the Guidelines of the GRI in the most precise manner, including as proof of this the table of correspondence among the indicators used and that provided for in the GRI Model, along with the Global Compact principles, Principles for Responsible Management Education and UI GreenMetric World University Ranking 2011e ISCN (International Sustainable Campus Network).

Once the reference guidelines had been identified, the work group analysed and selected the indicators to use, making this choice preliminary to ensure proper collection and assessment of the specific information. Whereas no particular problems were recognized with the data retrieval from an economic standpoint, it being possible that these would be taken from financial documents that are compulsory by law, the collection of information concerning the environmental dimension and the social dimension proved to be especially complex. Ca' Foscari did not

DOI: 10.1057/9781137351937

have an internal information system in control of the collection and the systematization of all the data of a social and environmental nature concerning the university. Such data was located, often in a fragmented manner, in various organizational phases: the collection of the information, its systematization and interpretation from a global point of view, with relevant analysis of the impacts, proved to be especially complex.

Concomitantly to the data collection, the university proceeded with the mapping of internal and external stakeholders, according to the modalities illustrated previously. In order to write the report, a series of 'interviews' of department heads were carried out, in order to share with them the identification of the indicators to measure and represent the economic, social and environmental impact in the departments. These indicators and background data are fundamental to the monitoring, controlling and reporting of the overall performance of the organization and its progress over time; the choice of the information was validated through a comparison between the selection and the requirements of the GRI. The university's choice was therefore that of a participatory approach, actively involving stakeholders in the decision-making processes (AccountAbility et al., 2005b).

After collection, the analysis of the data began, with reference to the objectives indicated in the sustainability policy adopted by the university, in order to combine both the planning and the controlling perspectives in the report.

The subsequent phase was that of writing the report, which was done by the Special Processes and Projects Office. It is a step with high criticality: the choice of the sections in which to organize the document and the space given to each one, and the modalities of presentation of the information, constitute essential elements that allows the disclosure and appraisal of the presentation of the sustainability strategies.

Ca' Foscari's report is organized in four sections: the institutional point of view, the social point of view, the environmental point of view and the economic-financial point of view.

In the first section, the institutional point of view, the communication (a Letter) with which the Rector, according to the provisions of GRI in the context of the sustainability principle (GRI, 2006), opens the document, contextualizing it with the university's sustainability policy, explaining the university's commitment to it, together with the main elements of innovation at the strategic and organizational level coming from acceptance of the sustainability point of view. Through the

DOI: 10.1057/9781137351937

Rector's contribution the will to integrate the sustainability concept in a widespread manner throughout the organization is expressed clearly (Bonn and Fisher, 2011), promoting awareness and understanding of the actions directed to that end, as the environmental sustainability and the creation of a fair society (Clugston and Calder, 1999). The first section contains, subsequently, the university's values, mission, vision, financial highlights and Key Performance Indicators, the sustainability policy, with the Sustainability Commitments Charter and its evolution, the stakeholder engagement, the organizational structure, the governance system focusing on the educational offer and research projects relevant to the sustainability theme. All of these are core issues of the university as an institution, and therefore Ca' Foscari pays a lot of attention to them.

The second section is dedicated to the social point of view and presents, through a qualitative and quantitative report, Ca' Foscari's actions in favour of the internal and external stakeholders, divided into students, human resources, community and territory. The third section concentrates on the environmental point of view, with special reference to the main repercussions that the university's activities generate on the ecosystem. The environmental theme is broken down into the supply chain, energy, water, materials, waste and mobility. Among the various initiatives, the already cited projects of the Carbon Management Project and the Sustainability Report were pointed out, as important constituent elements in the general picture, in strategic-organizational terms, an approach that has been adopted by Ca' Foscari. From the perspective of a sustainable supply chain (which consistently acknowledges a sustainability concept that requires an approach that ignores the boundaries of the organization) the organization of a Sustainable Public Procurement (SPP) system is illustrated. This system relies on the integration of environmental and social criteria, in addition to the economic and technical ones, criteria that are used in the choice of the suppliers of goods and services. Moreover, the means of monitoring, measuring and rationalization in the use of resources, for example energy, water and various materials, are disclosed. Furthermore, another critical objective in environmental terms is presented, the sustainable mobility policy adopted by the university and the relevant actions introduced by the mobility manager.

The fourth and last section deals with the economic-financial point of view, illustrating, among other aspects, the main sources of the funding, the comparison between average income per student and average cost

DOI: 10.1057/9781137351937

of teaching per student, the investments dedicated to sustainability, the economic impact on the community, the criteria for the selection of suppliers, and the university's participation in other organizations (societies, associations, syndicates and foundations).

The complete version of the report is a document with the four constituent sections substantially equivalent in terms of dimension, to attest to the equal importance given to each point of view. In fact the economic point of view has been slightly reduced compared with the others, a choice ascribable essentially to the presence of other reports with an economic and financial nature (provided for by law), while the social and environmental perspectives have the main part of the document.

With the second edition of the Sustainability Report (Ca' Foscari University, 2012c), the choices made in the first edition were confirmed, particularly the ones relevant to the contents and to the structure of the document. The assertion of the university's sustainability as 'strategic value' and an 'inspiring principle of Ca' Foscari', as well as 'wide ranging and foward-looking' (Ca' Foscari University, 2012c, p. 4) confirms again the acceptance of the sustainability paradigm amongst the pillars of the strategy, since 'having the awareness of one's own habits and their effect on the environment is the first step towards that mechanism of actions from below really and truly capable of modifying the situation' (Ca' Foscari University, 2012c, p. 4).

The 2011 report, of a slightly larger dimension than the previous one, maintains the subdivision among the four points of view, giving particular emphasis, among various aspects, to the participation of the university to the international network relevant to sustainability, to the results achieved in the Carbon Management Project and in the dematerialization process. The already cited Carbon Footprint Calculator is also presented, as an interactive tool for 'the awakening to and the promotion of the awareness of sustainability inside Ca' Foscari's community' (Ca' Foscari University 2012c, p. 82). The total space dedicated to the economic-financial point of view is further reduced, affecting the contents of the quantitative information, whereas the rest of the space dedicated to the economic impact on the community is unaltered. An element of obvious evolution of the report is the table of GRI correspondences, in an appendix, to attest to the greater adherence to the GRI Guidelines and the tendency towards a standardization of the contents, also in the interest of greater comparability of the performance over time and

DOI: 10.1057/9781137351937

among various organizations, as well as to improve the level of clarity and comprehensibility of the document for the stakeholders.

In terms of contents, another aspect of high importance is that the 2011 report considered the stakeholder engamenent theme. In addition, a revision of the mapping proposed the previous year was made as well as another element of strong innovation concerning the modalities of involvement of the stakeholders.

This activity proved to be challenging for the Office and for the whole organization.

Initially, the revision of the table used for the mapping of stakeholders was proceeded with, analysing the various actions of stakeholder engagement (for example, the possibility of receiving external inputs to the Sustainable Ca' Foscari site and the Rector's blog) with a view to pointing out the target stakeholders and the positive and negative repercussions towards stakeholders.

With special reference to the latter, three fundamental parameters on which the effectiveness of the actions of involvement introduced by the university depend were identified, namely:

- Frequency: the periodicity with which the involvement is activated by Ca' Foscari;
- Intensity: the force that the tool has in the involvement of the stakeholder considered;
- Importance: the importance of the body or of the subject involved.

A score (numeric figure) was assigned to each parameter on a fixed scale, which provided an overall ranking, following a rigid procedure.

Finally, for each action, the motivation that drove the university to promote it was expressed clearly, making reference to four alternatives: customer satisfaction, institutional, partnership, compulsory by statute.

Figure 3.8, by way of an example, is a graphic representation of the synthesis presented in the 2011 report for the category Firms.

The results of the analysis point out that the tools of stakeholder engagement reach on the average a good ranking. Among the various aspects that surfaced, the Rector's blog proved to be a particularly valid methodology for the internal stakeholders, while less effective were the tools of 'mass communication', such as social networks, the magazine *Ca' Foscari* and some types of events (Spritz Ca' Foscari and Ca' Foscari International Lectures). This is because, even though a high number of stakeholders were involved, both internal and external, these tools are

DOI: 10.1057/9781137351937

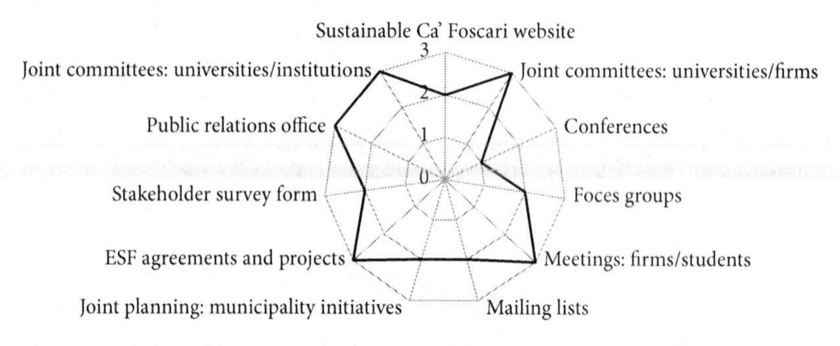

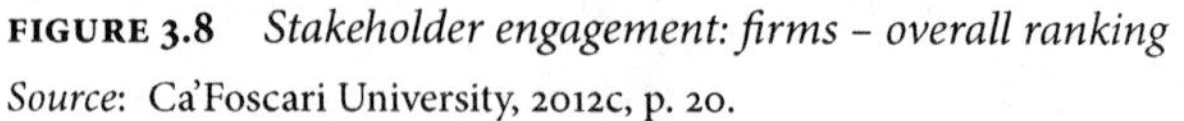
FIGURE 3.8 *Stakeholder engagement: firms – overall ranking*
Source: Ca'Foscari University, 2012c, p. 20.

mainly of a generic nature and do not stimulate the active participation of either the subjects involved or the employees.

The final step in the reporting process, once the writing of the document has been completed and approval obtained from the governing bodies, is publication. The circulation of the report to the public constitutes a crucial step in the process of stakeholder engagement. With full awareness of this, Ca' Foscari adopted various tools for the communication to stakeholders of the news of the publication, from the institutional newsletter to the university internet site, in addition to notices posted in the various premises of the university, and presented the report in the course of a public conference, with the participation of the Rector, of the Representative of the Rector and of the representatives of some of the major groups of stakeholders, including students, state and private businesses, and institutions.

The report was published in two versions: a concise version, the only one available both in digital and in paper format, and an unabridged integral version, available only on line. The choice is as a result of the various stakeholders, which prefer to communicate in different ways; at the same time it had a positive effect on the reduction of paper by avoiding a lot of printed reports.

If, on the one hand, publication represents a closing moment, in substantial terms the report represents one of the phases in the broader governance process of sustainability. According to the approach adopted by Ca' Foscari, the Sustainability Report does not constitute a seperate event, so that sustainability manages to permeate the daily processes of the organization, acquiring greater credibility

DOI: 10.1057/9781137351937

and legitimation. The subsequent constant monitoring of the results of the activities presented in the report is fundamental, together with the feedback coming from the process of stakeholder engagement. Ca' Foscari periodically presents the main outcomes of the monitoring process and relevant information to stakeholders, above all those of an environmental and social nature, through intermediate reports available on line. The feedback from the stakeholders fuels and enriches the analysis process and the identification of the appropriate corrective actions; at the same time, theis feedback helps to improve the same sustainability report, satisfying the information needs expressed by stakeholders.

Ca' Foscari University's objective is to throw itself into the challenge of integrated reporting, 'a process that results in communication, most visibly a periodic "integrated report", about value creation over time. An integrated report is a concise communication about how an organization's strategy, governance, performance and prospects lead to the creation of value over the short, medium and long term' (IIRC, 2013). Writing an integrated report is much more than simply making the various reports converge into a single document; it is relevant to the acceptance of new principles and new logics of building the document: strategic focus, connectivity of information, future orientation, responsiveness of stakeholder inclusiveness, conciseness, reliability and materiality (IIRC, 2013).

Aware of the extreme complexity of the challenge it faces in the decision to move towards integrated reporting, the university has activated a process aimed at controlling the conceptual pillars, desiged to generate impact on the strategic, management and organizational front, with particular emphasis on the theme of stakeholder engagement. The 2012 version of the report therefore contained elements of innovation like the two that preceded it, moving towards the challenge of an integrated report, which presumably will become a future trend in reporting.

Another element of innovation that Ca' Foscari intends to activate with the 2012 report is an interactive web platform, which represents a further way of disclosing the contents of the report to stakeholders, giving the reader the possibility not only to view the contents, but also to interact, screening them according to the 'objects' considered relevant, for example, extracting partial information reports, by sustainability dimension (environmental, social, economic) or whatever.

DOI: 10.1057/9781137351937

3.3.4 Sustainability teaching and research

In Ca' Foscari's Strategic Plan, as pointed out previously, sustainability becomes one of the fundamental axes of Ca' Foscari's path. In particular, in Objective 10, research and teaching assume a crucial role in the three-year planning period 2012–14 (Ca' Foscari University, 2012a, pp. 52–3); it is stated that Ca' Foscari aims 'to assume a transversal orientation of sustainability', becoming key in the 'teaching' and the 'development of the sustainability research' core issues. It is an extremely important undertaking, since didactics and research are the core issues of university institutions.

Through the formalization of dedicated strategic lines in a document which is the Strategic Plan, Ca' Foscari effectively recognizes explicity the importance of having a didactics offer and research activity greatly influenced by sustainability and proceeding with a strategic approach. The issue of sustainability is tackled in a formal manner, therefore not leaving it to the sensitivity of academic staff or researchers,but planning actions and pushing behaviours inspired by sustainability and making it strategic in nature.

In a first phase, the university was commited to the recognition and the improvement of the existing situation, with subsequent identification of the most effective tools to further strengthen the sustainability contents in the educational offer and in the research projects, both from the quantitative point of view and the qualitative one.

In particular, for the assessment of the existing situation, the identification of a series of keywords considered useful in order to carry out a search of Ca' Foscari's database and the national database (Uniservice) was proceeded with. In Figure 3.9 a graphic representation with the keywords adopted for the analysis of the existing situation is presented.

With regard to teaching, in order to strengthen the sustainability themes in the teaching programmes, a detailed analysis of the educational offer for the academic year 2011–12 was carried out, to identify the contents of sustainability embedded in various teaching, even if partially and not for the whole course. Furthermore, an analysis was carried out of the modalities of giving the lectures, aiming at identifying the courses that are given physically or using sustainable modalities, that is with modalities that allow a reduction in the use of paper resources in favour of online interactive tools and of open-source software. This first analysis, regarding the academic year 2011–12, was carried out by the

DOI: 10.1057/9781137351937

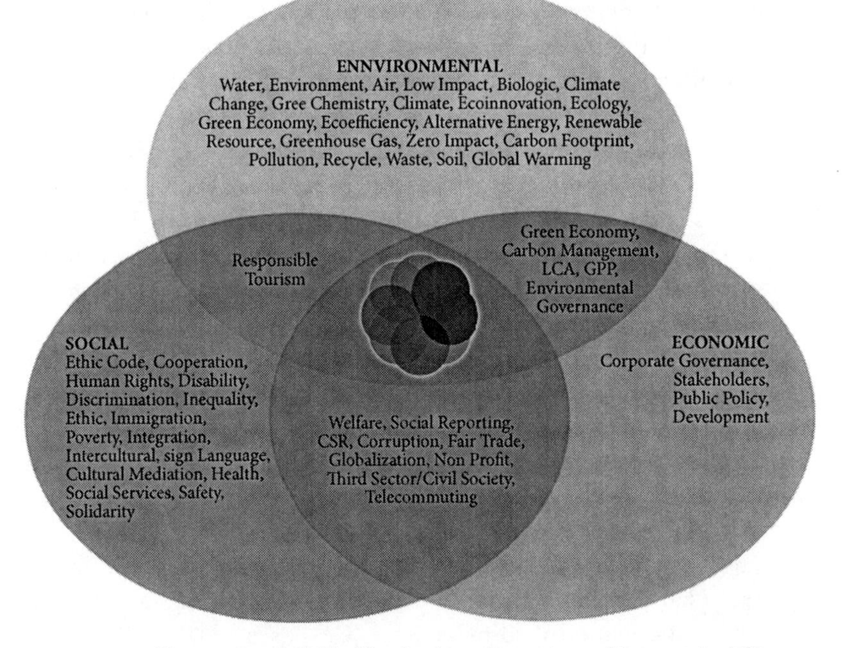

FIGURE 3.9 *Key words subdivided by the three dimensions of the sustainability*
Source: Ca' Foscari Venezia, 2013e.

departments and the Interdepartmental Schools, while for the following academic year (2012–13) the direct involvement of the professors was measured directly by professors, while they have been asked of filling in the teaching file, they must report if aspects of sustainability are present in their courses, in the contents and in the teaching methodology.

As a next step, the university will define guidelines that determine in a systematic manner the courses that, directly or indirectly, are oriented to the education and training of environmental and social sustainability.

The final aim of this line of action is that of enhancing the sustainable educational paths that the university makes available, so that this can become a competitive advantage that hopefully will be considered by the students, enabling them to construct their university careers on the basis of the sustainability point of view.

In the measurement of sustainability didactics, the path followed by Ca' Foscari has thus followed a first phase in which it proceeded in, retrospect, reconstructing and mapping the existing situation, preparatory

DOI: 10.1057/9781137351937

phase for the planning prospect, to be adopted in the following academic years. Indeed, the process of setting teaching objectives with 'sustainability' contents was started in order to sensitize first of all the teachers and also to encourage professors with different backgrounds to adopt the sustainability strategy and paradigm. To 'enhance the Sustainable Education Offer already activated, through the special web section and planning of new sustainable educational path' (Ca' Foscari University, 2012a, p. 54), the indicator identified by the university, and put in the Strategic Plan, is the relationship between the number of CFU ascribable to sustainability and the total number of CFU dispensed in the year.

On the research side, Ca' Foscari has activated a series of tools for its enhancement. One relevant contribution came from the increasing awareness of the importance of measurement and competition. In addition a tool such as the new 'UGOV catalogue' of the university's research has been filled in with increasing attention, and which tracks the sustainability profile.

In the case of Ca' Foscari University, the indicators identified in the Strategic Plan for the 'increase and the enhancement of the Research Projects in the matter of sustainability' (Ca' Foscari University, 2012a, p. 54) are the average between:

- Themed projects (PRIN, FIRB, UE) presented by professors and researchers and total projects presented;
- Financial resources acquired for themed projects and total resources acquired for research projects;
- The number of themed publications and the total number of publications.

Another tool is the development of the 'mapping of expertise', a web application that helps in the classification of the types of research and teaching expertise present in the university, including those pertaining to sustainability. It joins the other tools devoted to students, all of them targeted at shaping citizenship and education in the long term, and it is extremely powerful, above all in the synergies that can develop with other tools activated by Ca' Foscari, in particular with the Carbon Footprint Calculator. This is because it enables students to trigger and strengthen the engagement process not only from the quantitative point of view, in involving a high number of subjects, but also at the level of its transversality, since it reaches students belonging to all the disciplines.

DOI: 10.1057/9781137351937

The reorganization of the Uniservice website, dedicated to research and development, joins these tools, and in the future it will allow the publication in a user-friendly manner of the information relevant to products, projects, research groups and people present in the university.

Finally, another extremely powerful tool at the level of the dissemination and internalization of the sustainability principles, for the training of 'responsible citizens' who will adopt sustainable life styles, is 'sustainability expertise'.

The acquisition of sustainability competences (CDS) is tied to the realization of voluntary activities. It is an opportunity for every Ca' Foscari student at every level, and allows them to get one extracurricular ECTS (European Credit Transfer System), which will count in their university career and will be reported in their final certificate of exams taken. The students that achieve the ECTS on CDS are mentioned in a special way during the ceremony of proclamation.

In every department at least one representative is available with whom the interested student can agree on the theme to tackle and the modalities of execution. Among the thematic possibilities are the following: sustainable mobility, sustainable energy, sustainability reporting, historic analysis of the economic, social and environmental sustainability models, use of renewable resources, use and management of water resources, use and management of energy resources, analysis of the energy and material flows, anthropocentrism and geopolitics in sustainability, linguistic and cultural activities on sustainability, sustainability and the travelling experience. The activities can consist, according to the theme chosen, in attending seminars and workshops, carrying out bibliographical and scientific studies, activities in the field and the like. These activities can be carried out by students individually or in groups, perhaps with interdepartmental collaboration, with a final phase that involves the production of a paper, subject to evaluation by the representative, according to generally agreed criteria. The final paper can be realized in different ways, from the traditional 'short dissertation' to a video or a journalistic article, according to the activity carried out and however coherent with the workload foreseen and are couted as 1 ECTS. The papers produced are collected every six months in a digital review, and contribute to a special database on sustainability expertise, which can be freely consulted.

This tool adopted by Ca' Foscari is intended to pursue, once again, objectives of an educational nature, to stimulate in-depth analysis of

DOI: 10.1057/9781137351937

sustainability themes broken down into numerous components, the internalization of a sustainability culture and the embedding of its principles. At present, 1,000 students have chosen to undertake the activities necessary to acquire sustainability expertise, an extremely positive result and proof of the effectiveness of the approach to stakeholder engagement adopted by the university.

3.3.5 Other projects and intiatives

The other projects and initiatives started in the last few years by Ca' Foscari on its path towards sustainability have heterogeneous characteristics, albeit with the common denominator of the strategic-organizational approach adopted by the university. These other projects and initiatives do not therefore represent separate, one-off actions, but they fit into the sustainability framework structured by the university, apart from being targeted at achieving concrete results, even of a more immediate nature with regard to the relationship with the environment and society.

The already mentioned Selective Collection of Waste at Ca' Foscari project, a collaboration with the firm responsible for the management of waste in the Venetian province (Veritas S.p.A.), started in an experimental manner in 2010 in the pilot sites of central Ca' Foscari and the campus of S. Giobbe. The following year the project was extended to other buildings of the university, finally including all the premises of Ca' Foscari in Venice and in Mestre in 2012. The project involves the distribution of cardboard bins for the selective collection of paper, glass and plastic.

The aims pursued by Ca' Foscari with this project, one of the first important steps on the path towards sustainability, are essentially twofold and complementary. On the one hand, the project aims at achieving improvements in the rationalization of the system of selective waste collection at the University's premises, with consequent reassessment of the whole process. This objective is translated into the predisposition of the operational protocols within the municipal firm, aimed at integrating the process of waste collection into the day-to-day activities of the university. The protocols have been addressed also to companies that are responsible for cleaning and for the caretakers' lodges, supported by the supply of a specific training activity tailored for them. This is the perspective of answering effectively the expectations and the requirements, first of all, of students and staff, rather than, more generally, the people with various titles who attend to the Ca' Foscari structures. The university has

DOI: 10.1057/9781137351937

also activated a monitoring system of the activities carried out, in order to assess the progress of the selective collection of waste in each branch of the university, identifying also the complexity of each type of waste, to seize any criticalities and provide the relevant improvement actions.

On the other hand, this project pursues a fundamental educational aim, of provoking, promoting and encouraging sustainable behaviour among the Ca' Foscari community. It has special regard for the internalization of the principles of environmental sustainability in the actions carried out daily, which are, in this specific case, the production of waste, its handling with separation modalities to reduce the negative impact on the environment, at the beginning of the process and to coherently orient behaviour and purchasing decisions and the conscious use of goods and resources.

In this sense, numerous awareness initiatives directed above all to students and staff were activated. They included, besides the already mentioned training of the staff of the cleaning firm in the modalities of the university's waste collection system, participation in the European Week for Waste Reduction (EWWR) in 2010, 2011 and 2012, a three-year project supported by the European Commission through the LIFE+ programme, which comprises a series of events of the national and local levels, run by authorities, institutions, administrations, associations, non-profit organizations, schools, universities and businesses. Other activities involve students in a direct manner, as in the case of the Differenziatest, a prize contest aimed at the modalities of collection and of seperation of waste, which saw vast participation by the students of Ca' Foscari, with 369 questionnaires filled in in the ten opening days of the contest. The questions the students were asked related, for example, to the identification of the type of waste to which a specific object belongs (lighters, CD/DVD cases, clean plastic plates, dirty plastic glasses, greaseproof paper, drink cartons, receipts, etc.) and the correct behaviour in certain situations (for example, what to do after drinking fruit juice from a carton). In evaluating the answers not only the correctness or the difficulty of the questions was considered, but also the gravity of the wrong answers in terms of the environmental management of the waste cycle.

Another initiative aimed at students, which will be in its third edition in 2013, is the Ca' Foscari Short Film Festival. It is an event that has a peculiarity and a strong characteristic. It is oriented to the future, with students who are the real protagonists on both sides of the screen. The initiative consists in the staging of short films that are then selected and

DOI: 10.1057/9781137351937

judged by an international jury. The films are made not only by students of cinema schools, but also by students of courses in communication science, students from all over the world. The novelty of the 2013 festival, added precisely to point out this attention to the future and to the new generations, which constitute one of the fundamental axes of the initiative, is the contest 'Short&Sustainability', dedicated to short films that have sustainability as their theme, in its three constituent dimensions: economic, social and environmental.

The Climate Change Contest 2013 is another initiative, promoted by the Shylock Theatre University Centre Association of Venice and the Comete web magazine, with sponsorship by the Italian National UNESCO Committee and the Ca' Foscari University and with the collaboration of Sustainable Ca' Foscari. It is a national literary contest, conceived and promoted to develop and stimulate creativity and reflection on the theme of climate change, which is becoming every day more real and around which the projects are multiplying, with a significant increase in public awareness of the matter. With the 2013 edition the modalities of participation and artistic means admitted were improved, precisely with the aim to stimulate a greater number of people to express their ideas on the theme in discussion. Specifically, in addition to the category 'Fiction and image' the category 'Web gallery' is present. This section, developed in collaboration with Sustainable Ca' Foscari, was expressly made with an orientation towards the language of the social media, to stimulate the participation of the young and students. The category 'Fiction and images' is organized in three sections (two for fiction and one for images) and the work submitted is judged by a jury of experts, comprising journalists, editors and heads of newspapers and of programmes that deal with environmental themes.

On the theme of energy conservation, which Ca' Foscari, as pointed out previously, pays great attention to, for example, the impacts in the Carbon Management Project, 'I light up less' (M'illumino di meno) is an initiative that has already become a 'habit' for the Venetian university. In general, the initiatives that are repeated from year to year have innovative elements introduced in each edition, with the purpose of increasing their effectiveness and diffusion. The 'energy conservation party' fits into the scope of the initiatives of an educational nature, aiming at making people acquire, specifically, the awareness that energy conservation requires daily attention and that small actions carried out by everyone are fundamental to achieving results on a large scale. For the 2013 edition Ca' Foscari has

DOI: 10.1057/9781137351937

proposed various actions, precisely in order to involve a high number of stakeholders, from the technical staff to the teachers and students, so that everyone inside the university may express their support. Once again the seal of approval of the leaders was fundamental to the spreading of information concerning the initiatives and activities organized by Ca' Foscari during the year, to reduce waste as much as possible and improve the energy efficiency of the university. The actions realized in the last edition express, even with different values, modalities and contexts, the great attention that the university and its stakeholders, by their participation, gave to the theme. Among the various initiatives were:

- The lesson by candlelight, during the day of Friday 15 February 2013, when many professors chose to do a quarter of an hour of the lesson by candlelight, using the candles distributed across the various sites of the university, to sensitize students to energy conservation. In some cases the professors decided to concentrate the whole lesson on themes tied to the efficient use of energy. Some testimonies of this experience were then published through the social networks (in the Twitter profile and on the Facebook page of Sustainable Ca' Foscari);
- The flash-mob in San Sebastiano, organized by the students of the course in interior architecture and organization together with the professor, preparing the site of San Sebastiano with about 600 candles and then reaching the headquarters of Ca' Foscari, linking the two sites symbolically with these candles and then, switching off both sites at the same time for 'energy silence';
- The switching off of all the lights in Ca' Foscari's central site for an hour and the lighting of the same site with LED lights;
- The distribution of bookmarks in the days preceding the initiatives at the various sites of the university on which specific suggestions on the modalities of waste reduction and on the behaviour to adopt to become more sustainable university students were explained.

'The Garbage Patch State' is another project recently started by Ca' Foscari. This concerns the environmental dimension of sustainability and fits within the proposals for activities that permit the acquisition of sustainability expertise, as previously mentioned. This project was both educational and provocative concerning the extensive patches of waste deposits in the oceans, whose size increases from year to year. In the first phases of the

DOI: 10.1057/9781137351937

project the artist that started the initiative, supported by the collaboration of Sustainable Ca' Foscari, decided to involve all the students interested in sustainability themes, giving them the opportunity to express their own creativity through their participation in the project. The contributions of the students were collected on the site created expressly to promote the project. The output of the project, which becomes an original art installation at Ca' Foscari, is an installation in the courtyard of the central site, on the occasion of the 55th Art Biennale. The project was officially launched at the UNESCO offices in Paris in the month of April 2013 and the installation is to be displayed at Ca' Foscari from 1 June to 24 November 2013.

In the field of sociality, a project activated by the university is 'Social Ca' Foscari', a collaboration with public organizations and non-profit associations of Veneto aimed at developing the knowledge, expertise and capabilities of the Ca' Foscari community and facilitating meetings between members of the university and local people involved in voluntary work, for a greater understanding of their actual needs. The objective of this project is to further increase awareness of the social role of the university, concomitantly making the most of the professionalities and capabilities of the individuals who demonstrate sensitivity and willingness to lend their activities to this goal. The university aims to become part of the dynamics of territorial welfare, precisely through its co-operation with non-profit associations, a co-operation that is made possible by the use of the voluntary work of the employees of the university. At the moment the project is aimed at the academic and the technical-administrative staff of the university, but the aim for the future is to extend it to other individuals, including students. In this project the university provides its support; in actual fact the Ca' Foscari brand is present in the collaboration project. Among the distinguishing characteristics of 'Social Ca' Foscari', especially significant is the fact that participation is voluntary and therefore no gain or wages are expected, with activities to carry out outside work hours (paid by the university, financed with public funds). Furthermore, the endorsement of the university does not involve any useful recognition for employees, being this initiative a 'pro bono activity'. The success of the project relies on the awareness that the individuals who form the Ca' Foscari community are moved by values and an ethical approach inspired by responsibility towards the community where they live. Besides it is an opportunity to develop their professionalism and all the other capabilities, that in the day by day routines could also remain hidden or sleeping.

DOI: 10.1057/9781137351937

Particularly significant is Ca' Foscari's commitment to promote the integration of sustainability indicators in university rankings. This is a prospect that would consequently see sustainibility rise, in particular the environmental and social themes. The initiative 'Sustainability embedded in university rankings' aims at including sustainability indicators in order to spread a sustainable perspective into all academic institutions. Universities should integrate sustainability in their organisations and educational programs to build generations of sustainable leaders and contribute to the creation and spread of a culture that underpins sustainable behaviour. There are various rankings that specialize in sustainability, but an integration of sustainable indicators into traditional rankings will stimulate a stronger awareness of the issue and consequently the inclusion of sustainability into university agendas. To launch this process it is necessary to share idea and motivations with relevant and authoritative players of the rating systems and with experts in sustainability and sustainability rankings. The success of this project will be increased through the selection of key players engaged in this initiative, and of those who share a common motivation.

The operationalprocess to realize this project is structured in these two fundamental steps:

1. Ca' Foscari University organises a workshop in Venice in October 2013 to share with others the aim of including sustainability in university rankings. After this event a protocol would be released, to share goals and commitments towards the objective and stating roles for each player. On this occasion a round table of technical experts would be established to discuss and select the set of indicators to include;
2. To take part in 'EAIE 2014 – European Association of International Education' and present a paper outlining the operational proposal to integrate sustainability into rankings.

FIGURE 3.10 *Sustainability embedded into university rankings*

DOI: 10.1057/9781137351937

Appendix: Guidelines for Carbon Management in Italian Universities

The Guidelines for Carbon Management in Italian Universities were developed jointly by the Ministry of the Environment and the Protection of the Territory and of the Sea (MATTM) and Ca' Foscari University Venice (UCF).

They are the first product of this collaboration and of the agreement between the ministry and the university on carbon management themes. The sustainability strategy adopted by the Venetian university provides for, among other actions aimed at improving the social role and reducing the environmental impacts of the university, a Carbon Management Project, an initiative of a highly innovative nature in the specific context of higher education. Carbon management, defined in the project as the core management process, was included in the Sustainability Commitments Charter, the first official document of fundamental importance for Ca' Foscari in terms of the public assumption of responsibility in a medium- to long-term perspective.

The Ca' Foscari initiative is in step with the one already started by the Ministry of the Environment and the Protection of the Territory and of the Sea for the fostering of agreements with public institutions and private businesses, targeted at the reduction of greenhouse gas emissions, reductions that are realized through specific projects connected to the calculation of the carbon footprint. The project of the Venetian university represents a pilot case of

DOI: 10.1057/9781137351937

the application of carbon management principles to a university structure in Italy and, among its aims, it includes the writing-up of guidelines based on a concrete study case.

Besides proposing an approach that enables the definition of a common emissions management policy for universities, the guidelines use the concrete results and the actions carried out by Ca' Foscari University Venice in the execution of the project; the initial phase of the project was supposed to end in September 2011. The definition of the guidelines is a combination of strategic procedural and organizational aspects on the one hand, technical-methodological aspects on the other. The underlying objective of these guidelines is that of pointing out the unavoidability on the part of the university of a widespread and organic approach, which derives from a general perspective, strategically oriented towards sustainability.

Ca' Foscari obtained ISO 14064:2006 certification for the years 2009 and 2010 in the scope of the Addendum to the Carbon Management agreement entered into with the Ministry of the Environment, which renewed the collaboration. Bureau Veritas, an external certifying body, duly attested that the calculation of the university's carbon footprint confirms to the international standards for environmental management.

The guidelines are a reference document not only for the Italian universities that may want to develop their own carbon management, but in general also for public and private education institutions.

DOI: 10.1057/9781137351937

Bibliography

AccountAbility AA1000 (2011), *Stakeholder Enagagement Standard 2011: Final Exposure Draft*, London: AccountAbility.

AccountAbility and United Nations Environment Programme Stakeholder Research Approach (2005a), *The Stakeholder Engagement Manual. Volume 1: The Guide to Practitioners' Perspective on Stakeholder Engagement*, Cobourg, Ontario: Stakeholder Research Associates Canada Inc.

AccountAbility and United Nations Environment Programme Stakeholder Research Approach (2005b), *The Stakeholder Engagement Manual. Volume 2: The Practitioners' Handbook on Stakeholder Engagement*, Cobourg, Ontario: Stakeholder Research Associates Canada Inc.

Albrecht P., Burandt S. and Schaltegger S. (2007), Do sustainability projects stimulate organizational learning in universities?, *International Journal of Sustainability in Higher Education*, 8(4), 403–15.

Ballou B., Heitger D.L. and Landes C.E. (2006), The future of corporate sustainability reporting: a rapidly growing assurance opportunity, *Journal of Accountancy*, 12, 65–74.

Bebbington J., Larrinaga C. and Moneva J.M. (2008), Corporate social responsibility reporting and reputation risk management, *Accounting, Auditing & Accountability Journal*, 21(3), 337–61.

Becher T. and Kogan M. (1992), *Process and Structure in Higher Education*, London: Routledge.

DOI: 10.1057/9781137351937

Bergamin Barbato M. and Mio C. (2008), Il bilancio socio-ambientale nei processi di innovazione delle aziende e delle amministrazioni pubbliche [Social-environmental report on innovation processes of public administrations], in Gruppo di studio e attenzione dell'Aidea (ed.) *Innovazione e Accountability nella pubblica amministrazione. I drivers del cambiamento* [Innovation and accountability in public administration. The drivers of the change], Rome: RIREA, pp. 154–238.

Bonn I. and Fisher J. (2011), Sustainability: the missing ingredient in strategy, *Journal of Business Strategy*, 32(1), 5–14.

Calder W. and Clugston R.M. (2003), International efforts to promote higher education for sustainable development, *Planning for Higher Education*, 31(3), 34–48.

Clugston R.M. and Calder W. (1999), Critical dimensions of sustainability in higher education, in Filho W.L. (ed.), *Sustainability and University Life*, Frankfurt: Peter Lang.

Cortese A.D. (2003), The critical role of higher education in creating a sustainable future, *Planning for Higher Education*, 31(3), 15–22.

Crane A. and Matten D. (2004), *Business Ethics: An* [sic] *European Perspective*, Oxford: Oxford University Press.

Daub C.H. (2007), Assessing the quality of sustainability reporting: An alternative methodological approach, *Journal of Cleaner Production*, 15, 75–85.

Fonseca A., MacDonald A., Dandy E. and Valenti P. (2011), The state of sustainability reporting at Canadian universities, *International Journal of Sustainability in Higher Education*, 12(1), 22–40.

Gadsby H. and Bullivant A. (2010), *Global Learning and Sustainable Development*, London: Routledge.

Gesualdi F. (2005), *Sobrietà: Dallo Spreco dei Pochi ai Diritti Per Tutti* [Moderation: from waste of few people to rights for everyone] Milano: Feltrinelli.

Gray R. (2010), Is accounting for sustainability actually accounting for sustainability... and how would we know? An exploration of narratives of organisations [sic] and the planet, *Accounting, Organizations and Society*, 35, 47–62.

Gray R. and Milne M.J. (2002), Sustainability reporting: who's kidding whom?, *Chartered Accountants Journal of New Zealand*, 81(6), 66–70.

Hinna L. (2002), *Il bilancio sociale. Scenari, settori e valenze; modelli di rendicontazione sociale, gestione responsabile e sviluppo sostenibile;*

esperienze europee e casi italiani [Social report. Scenarios, industries and values: patterns of social reporting, responsible management and sustainable development: European experiences and italian cases], Milano: Il Sole24ore.

Jabnoun N. (2009), Economic and cultural factors affecting university excellence, *Quality Assurance in Education*, 17(4), 422–7.

Jones P., Selby D. and Sterling S. (2010), *Sustainability Education: Perspective and Practice across Higher Education*, London: Earthscan.

Kaplan R.S. (2001), Strategic performance management and management, *Nonprofit Organizations, Nonprofit Management and Leadership*, 11(3), 353–70.

Locatelli R. and Schena C. (2011), Responsabilità e rendicontazione sociale del sistema universitario: Il caso italiano [Responsibilities and social reporting of Italian universities: the Italian case], in Arcari, A. e Grasso, G. (eds), *Ripensare l'università. Un contributo interdisciplinare sulla legge n° 240 del 2010* [Rethinking the university. An interdisciplinary contribution on law 240/2010], Milan: Giuffrè.

Lozano R. (2006), A tool for a graphical assessment of sustainability in universities (GASU), *Journal of Cleaner Production*, 14, 963–72.

Lozano R. (2011), The state of sustainability reporting in universities, *International Journal of Sustainability in Higher Education*, 12(1), 67–78.

MacVaugh J. and Norton M. (2012), Introducing sustainability into business education context using active learning, *International Journal of Sustainability in Higher Education*, 13(1), 71–87.

Marshall R.S., Vaiman V., Napier N., Taylor S., Haslberer A. and Andersen T. (2010), The end of a 'period': sustainability and the questioning attitude, *Academy of Management Learning and Education*, 9(3), 477–87.

Ministry of the Environment and of the Protection of the Territory and of the Sea and Ca' Foscari University of Venice (2012), *Guidelines for Carbon Management in Italian Universities.*

Mio C. (2002), *Il Budget Ambientale. Programmazione e Controllo della Variabile Ambientale* [The environmental budget: planning and control of the environment], Milan: Egea.

Mio C. (2005), *Corporate Social Responsibility e Sistema di Controllo: Verso l'integrazione*, (Corporate social responsibility and managment control : towards integration) Milan: Franco Angeli.

Mio C. and Borgato B. (2012), *Performance Measurement nelle Istituzioni Universitarie: Verso Una Prospettiva Di Sostenibilità* [Performance

measurement at universities: towards a sustainable perspective], Rome: Rirea.
Musu I. (2009), Crescita sostenibile e innovazione ambientale: ruolo della regolazione e della responsabilità sociale [Sustainable growth and environmental innovation: the role of regualtion and social responsibility], *L'Industria*, 1, 59–72.
Rebora G. (2007), Ricerca senza qualità? Il caso delle scienze aziendali e del management [Research without quality? The case of economics science and management], *Liuc Papers, n. 209*, Serie Economia aziendale 31.
Sammalisto K. and Arvidsson K. (2005), Environmental management in Swedish higher education: directives, driving forces, hindrances, environmental aspects and environmental co-ordinators in Swedish universities, *International Journal of Sustainability in Higher Education*, 6(1), 18–35.
Savan B. and Sider D. (2003), Contrasting approaches to community-based research and a case study of community sustainability in Toronto, Canada, *Local Environment*, 8(3), 303–16.
Shriberg M. (2002), Institutional assessment tools for sustainability in higher education. Strengths, weaknesses, and implications for practice and theory, *International Journal of Sustainability in Higher Education*, 3(3), 254–70.
Stern N. (2006), *Stern Review on the Economics of Climate Change*, Cambridge: Cambridge University Press.
Trani E.P. and Holsworth R.D. (2010), *The Indispensabile University. Higher Education, Economic Development and the Knowledge Economy*, Plymouth, UK: Rowman & Litterfield.
Velazquez L., Munguia N., Platt A. and Taddei J. (2006), Sustainable university: what can be the matter?, *Journal of Cleaner Production*, 14, 810–19.
World Commission on Environment and Development (1987), *Our Common Future*, Oxford: Oxford University Press.

Webliography

AccountAbility (2013), *The AA1000 Standards*, www.accountability.org/standards/index.html, accessed April 2013.
Ca' Foscari University of Venice (2008), *Codice Etico dell'Università Ca' Foscari di Venezia* [The Ethical Code of Ca' Foscari University of

Venice], http://unive.it/media/allegato/regolamenti/codici/Codice-Etico-Vigente.pdf, accessed January 2013.

Ca' Foscari University of Venice (2011a), *Ca' Foscari: primi in Italia ad approvare il nuovo decreto* [Ca' Foscari: first in Italy to approve the new law], www.unive.it/nqcontent.cfm?a_id=87535, accessed March 2013.

Ca' Foscari University of Venice (2011b), *Sustainability Report 2010*, www.unive.it/nqcontent.cfm?a_id=135400, accessed March 2013.

Ca' Foscari University of Venice (2011c), *University Statute*, www.unive.it/media/allegato/ateneo/Statuto-Ateneo.pdf, accessed March 2013.

Ca' Foscari University of Venice (2012a), *Strategic Plan 3/2012. Towards Ca' Foscari 2018*, www.unive.it/media/allegato/comunicazione/inagurazioneaa/120309_Piano_Strategico_def.pdf, accessed March 2013.

Ca' Foscari University of Venice (2012b), *Sustainability Commitments Charter 2013–2015*, http://unive.it/media/allegato/sostenibilita-pdf/CIS_2013_2015.pdf, accessed March 2013.

Ca' Foscari University of Venice (2012c), *Sustainability Report 2011*, www.unive.it/nqcontent.cfm?a_id=135401, accessed March 2013.

Ca' Foscari University of Venice (2013a), *Ca' Foscari at a glance*, www.unive.it/nqcontent.cfm?a_id=120327, accessed March 2013.

Ca' Foscari University of Venice (2013b), *Welcome to Ca' Foscari*, www.unive.it/nqcontent.cfm?a_id=127406, accessed March 2013.

Ca' Foscari University of Venice (2013c), *Sustainability step by step*, www.unive.it/nqcontent.cfm?a_id=134324, accessed March 2013.

Ca' Foscari University of Venice (2013d), *Table of stakeholders*, www.unive.it/nqcontent.cfm?a_id=134335, accessed March 2013.

Ca' Foscari University of Venice (2013e), *Teaching & Research*, www.unive.it/nqcontent.cfm?a_id=132909, accessed April 2013.

CRUI Foundation (2010), *CAF Università. Migliorare un'organizzazione universitaria attraverso l'autovalutazione*, www.fondazionecrui.it/pubblicazioni/Documents/CAF_Universit%C3%A0.pdf, accessed January 2013.

CRUI Foundation (2012), *CAF Università. Il modello europeo di autovalutazione delle performance per le università* [The European self-assessment model for univeristy performance], www.fondazionecrui.it/pubblicazioni/Documents/CAF/caf_uni_2012.pdf, accessed January 2013.

European Institute of Public Administration (2013), *The Common Assessment Framework (CAF) 2013: Improving Public Organisations*

DOI: 10.1057/9781137351937

through Self-Assessment, www.eipa.eu/files/File/CAF/CAF_2013.pdf, accessed April 2013.

Global Reporting Initiative, GRI (2006), *Sustainability reporting guidelines G3*, https://www.globalreporting.org/resourcelibrary/G3-Guidelines-Incl-Technical-Protocol.pdf, accessed April 2013.

Global Reporting Initiative, GRI (2011), *Sustainability reporting guidelines G3.1*, https://www.globalreporting.org/resourcelibrary/G3.1-Guidelines-Incl-Technical-Protocol.pdf, accessed April 2013.

Global Reporting Initiative, GRI (2012), https://www.globalreporting.org/Information/about-gri/Pages/default.aspx, accessed December 2012.

International Integrated Reporting Council, IIRC (2013), www.theiirc.org, home page, accessed March 2013.

International Sustainable Campus Network, ISCN (2013), www.international-sustainable-campus-network.org, home page, accessed April 2013.

Latouche S. (2007), De-Growth: An electoral stake?, *International Journal of Inclusive Democracy*, 3(1), www.inclusivedemocracy.org/journal/vol3/vol3_no1_Latouche_degrowth.htm, accessed April 2013.

London Benchmarking Group, LBG (2013), www.lbg-online.net, home page, accessed April 2013.

Maddison A. (2005), *World Development and Outlook 1820–2030: Evidence Submitted to the House of Lords*, www.ggdc.net/maddison/oriindex.htm, accessed April 2013.

Principles for Responsible Management Education (2013), www.unprme.org, home page, accessed April 2013.

Schwarz J., Beloff B. and Beaver E., (2002), *Use Sustainability Metrics to Guide Decision-Making*, www.docstoc.com/docs/22957127/Use-Sustainability-Metrics-to-Guide-Decision-Making, accessed April 2013.

Social Accountability International (2013), *SA8000 Standard*, www.sa-intl.org/index.cfm?fuseaction=Page.ViewPage&pageId=937, accessed April 2013.

ULSF, University Leaders for a Sustainable Future (1990), *The Talloires Declaration. 10 Point Action Plan*, www.ulsf.org/talloires_declaration.html, accessed January 2013.

ULSF, University Leaders for a Sustainable Future (2013), *About ULSF*, www.ulsf.org/about.html, accessed April 2013.

DOI: 10.1057/9781137351937

United Nations (2002a), *Report of the World Summit on Sustainable Development*, www.johannesburgsummit.org/html/documents/summit_docs/131302_wssd_report_reissued.pdf, accessed April 2013.

United Nations (2002b), *What is the Johannesburg Summit 2002?*, www.un.org/jsummit/html/basic_info/basicinfo.html, accessed April 2013.

United Nations (2012), *Report of the United Nations Conference on Sustainable Development*, www.uncsd2012.org/content/documents/814UNCSD%20REPORT%20final%20revs.pdf, accessed April 2013.

United Nations Educational, Scientific and Cultural Organization, UNESCO (1998), *World Declaration on Higher Education for the 21st Century: Vision and Action*, www.unesco.org/education/educprog/wche/declaration_eng.htm#world%20declaration, accessed April 2013.

Universitas Indonesia, UI, Green Metric (2013), *UI GreenMetric World University Ranking*, http://greenmetric.ui.ac.id, accessed April 2013.

DOI: 10.1057/9781137351937

Index

DOI: 10.1057/9781137351937

DOI: 10.1057/9781137351937

CPSIA information can be obtained at www.ICGtesting.com
Printed in the USA
LVOW08*1222100813

347062LV00005B/8/P